Frequency

Chavan's UED research, Volume 3

Sandeep Chavan

Published by SANDEEP CHAVAN, 2025.

FREQUENCY

First edition. April 30, 2025.

ISBN: 979-8231951277

Written by Sandeep Chavan.

Table of Contents

Dedication

To the **scientists**, **researchers**, **students**, **science lovers**, and every silent **tuner of reality**—May your structures continue to align with deeper resonance.

And to **Max Planck**, **Heinrich Hertz**, and those who lived not in noise but in **frequency**—Your resonance reached us.This book is an attempt to complete the note you began.

Epigraph

"Light was never sent to you.It was always there—waiting for your resonance to realize it.You do not see the truth.You become the structure through which it appears."— *Chavan's Law of Realization*

Opening Note by the Author

We have been told that light travels. That it moves like a courier from source to surface. We have been taught that frequency is a fixed property—something embedded in the wave itself, measurable and absolute. We have believed that observation is passive—that we watch the universe unfold, untouched by our gaze.

But what if all of that is incomplete?

This book is not an argument—it is a tuning. It is not a theory to be proven—but a structure to be aligned with. It does not aim to redefine science by opposition, but to **reveal a deeper rhythm beneath it**—where reality is not delivered, but realized.

In these pages, we explore light without photons, vibration without transmission, and energy without particles. We explore **frequency** not as a number on a graph, but as **a signature of resonance between the field and the observer**. We ask: What is truly happening when we say we "see"? And more importantly: **Who is doing the seeing?**

Universal Energy Dynamics (UED), the framework that underpins this work, is not a replacement for physics. It is the **realization that physics itself is structured observation**, and that the observer is not external to truth—they are its condition.

So I invite you—not just to read, but to recalibrate. Let these chapters rearrange something in your thinking. Let them question not just what light is, but **what you are, as one who receives it**.

This is not a book about light. It is a book about *you*, tuning to the frequency of what has always been waiting to be seen.

— **Sandeep**

Preface

"You never saw light. You only saw your alignment with it."

Every time you turn on a bulb, glance at the stars, or stare at your phone screen, you think you're *seeing light*. But what if you're not? What if nothing is coming toward you at all—no photons, no particles, no waves—just a structured energy field waiting to resonate with your perception?

This book isn't here to glorify complexity. It's here to dismantle illusions that have quietly slipped into science and settled as untouchable truths. "Light travels," they say. "Energy comes in packets." "Frequency is an objective measure." We nod. We repeat. We build technologies on these assumptions.

But we never asked the deeper question: What is frequency, really—and what role do *we*, the observers, play in its very existence?

The Purpose of This Book

This book challenges the belief that frequency is a property of external phenomena. It proposes a fundamental shift:

Frequency is not an intrinsic trait—it is an emergent experience.

Our instruments don't measure absolute truth; they translate resonance. Our experiments don't reveal how the universe works; they reflect where our intent happens to align with universal structure.

The purpose of this book is to reclaim observation as the central act of realization, not passive reception. It's to show that all science—especially what we believe about light and energy—is born from conscious interaction with the universe.

Why Frequency Matters More Than We Think

We live in a world governed by frequency. It shapes our perception of color, sound, heat, and even the passage of time. But we rarely question what frequency truly is.

Is it a property of light? A characteristic of matter? A universal constant?

No.

It is a sign that alignment has occurred. It is the *footprint* of resonance between observer and vibration. Without the observer, frequency has no meaning—because without conscious time, it has no framework to emerge in.

Once we realize this, everything changes. The photoelectric effect, Planck's constant, even the idea of the photon—it all shifts. These are not hard truths of the cosmos. They are observer-bound interpretations of resonance.

From Observation to Realization

We haven't discovered the universe—we've only decoded those parts of it we were tuned to receive.

Science is not discovery. It is resonant interpretation. Equations like $E=h\nu$ are not truths—they are translations of our alignment into formal language. The photon is not a particle—it is a concept born out of mathematical necessity and observational convenience.

What we call "laws of physics" are the reverse-engineered consequences of our interaction with the field of universal energy. And the field doesn't speak in frequencies. It vibrates. Constantly. Quietly. Universally. Only when an observer tunes in, does frequency become real—and reality begins to glow.

This book is not about how the universe works. It is about how you come to know it—through resonance, intent, and alignment. You don't find the truth. You harmonize with it.

Welcome to *Frequency*—a new understanding of light, energy, observation, and intelligence itself.

How to Read This Book

This is not a textbook. It's not a manual. And it's definitely not a step-by-step guide to "understand the universe."

This book is a frequency map—a structure of insights, questions, and principles that resonate differently depending on where you are in your own thinking.

Here's how to approach it:

1. Read with Openness, Not Urgency

This book isn't designed to be rushed. It's meant to be felt as much as understood. Let your structure align naturally with the ideas. If something doesn't click today, it might tomorrow.

2. Treat Each Chapter as a Tuning Fork

Every chapter is a resonance space. Read it not just for facts, but for the shift it creates in your perception. Pause. Reflect. Revisit.

3. Don't Worry About Agreement

This is not about convincing you. This is about inviting you to think differently—to see familiar ideas through a new structural lens. You are free to question everything, including the foundations of this work.

4. Use the Diagrams and Glossary Liberally

Many concepts in this book move beyond language. The visuals and glossary are designed to help your intuition catch up to your logic. Don't skip them. Return to them often.

5. Pause at the Laws and Takeaways

Each Chavan's Law, reflection, and summary is a conceptual anchor. These are not just definitions—they are tools for perception. Let them recalibrate your thinking.

6. Skip Around If You Need To

The book flows linearly, but it's not fragile. If a later chapter calls to you, go there. If an appendix sparks interest, follow it.

This book honors structure—but **you are the tuning instrument.**

7. Let It Echo After You Close It

You won't "finish" this book. You'll tune with it. And then, like all resonance, it will continue shaping how you see, think, and align—with everything.

Welcome to the field. The signals are always there. This book is just one way to hear them.

Chapter 1: The Illusion of Light

"You never saw light. You only resonated with it."

Imagine a simple question: If light is always traveling through space, why do we only see it when we look?

Your room is filled with electromagnetic waves—visible, invisible, measurable. But the moment you close your eyes, where does the light go? The bulb may still be "on," but the phenomenon of light, as you know it, ceases to exist in your experience.

So here's the paradox: If light is something emitted—a packet or a wave— then it should be observable at all times, regardless of your awareness. But it's not. It's only there when you resonate with it—when you are tuned, attentive, and receptive.

This is not a philosophical illusion. It's a physical reality that has been buried under centuries of assumptions.

We have been told that light travels, that photons carry energy, that particles move across empty space and deliver brightness to your eyes.

But what if that's only half the story?

What if light is not something that moves toward you, but something you unlock through alignment?

What if the LED doesn't "emit" light in the traditional sense, but instead activates a structured energy field, waiting for you to tune in?

This is not just a different way of explaining light. It is a complete reframing of what we've believed about emission, energy, and observation itself.

Light is not something the universe sends. It is something your awareness resonates with.

And this simple shift—from emission to realization—will change how we understand not only light, but energy, perception, technology, and science itself.

Let's begin with the question we never truly asked: What is light—really?

1.1 What Is Light—Really?

We often use the word *light* as if it's an unquestionable truth—something obvious, direct, even self-explanatory. But if we peel back the layers of our assumptions, we'll find that the idea of light has been one of the most shape-shifting concepts in scientific history. It has evolved not as a stable entity—but as a moving metaphor, stretched by the limitations of human perception and rebuilt by the models of each era.

Let's step back for a moment.

From Newton to Now: The Evolution of an Illusion

In the 17th century, Isaac Newton imagined light as a stream of tiny corpuscles—particles ejected from luminous objects. His theory aligned well with the mechanics of his time, treating light like tiny projectiles that obeyed laws of motion. Reflection and refraction, in his view, were just directional changes of these mass-bearing light bits. It was a neat, mechanical explanation.

But then came Christiaan Huygens and later, James Clerk Maxwell, who offered a different model—light as a wave, not a particle. Maxwell's equations in the 19th century revealed that changing electric and magnetic fields could propagate through space in the form of electromagnetic waves, needing no medium. This explanation brought coherence to the behavior of light as seen in interference, diffraction, and polarization.

Yet, by the early 20th century, something didn't add up.

Phenomena like the photoelectric effect—where light could eject electrons from metal surfaces—couldn't be explained by wave theory alone. Light seemed to behave like it arrived in discrete packets of energy. And so Einstein stepped in, proposing the idea of the photon—a massless quantum of energy, traveling at light speed, carrying momentum, but no rest mass. The duality emerged: light was now both wave and particle, depending on how you looked at it.

Science accepted this paradox with mathematical convenience—but never truly resolved its conceptual contradiction.

But Is Light Really a Particle or a Wave?

We must now pause and ask: Are we describing what light is, or are we only modeling how it behaves in our presence?

If light is a particle, then what medium does it "travel" through when space is considered empty?

If light is a wave, where is the oscillating substance? What's waving in a vacuum?

If it's a field disturbance, how is that disturbance sustained across billions of light years without any medium to contain it?

Each model—particle, wave, field—has served as a scaffold for resonance, not a final declaration of truth. We've never discovered what light *is*. We've only described how it behaves when we observe it.

Introducing the UED Framing: Light as Structured Energy Seeking Balance

From the perspective of Universal Energy Dynamics (UED), light is not a substance or object—it is a patterned restructuring of energy fields in a medium that is never truly empty. What we perceive as "light" is the visible effect of energy imbalances trying to return to equilibrium through structured, directional disturbances.

- Light does not "travel" from point A to point B.
- Rather, energy restructures itself through space, and when an observer aligns with that structure—light is realized.
- There are no particles in flight. No waves moving "through" vacuum.
- There is only energy seeking balance, and perception arising from resonance.

This means that light is not emitted in the classical sense—it is activated, patterned, and recognized in the presence of tuned observers or systems.

The Photon as the Conceptual Pivot

This redefinition of light now demands a reevaluation of the photon itself.

- Is the photon a physical object, or simply a useful construct we created to match energy measurements with

mathematical formulas?
- Is the quantization real, or is it just the output of a resonant interaction between source, medium, and observer?

To understand this, we must now look more closely at the photon—not as a confirmed entity, but as the artifact of a specific interpretation of observation.

In the next section, we will challenge the reality of the photon itself, and explore how it was retrofitted into scientific thought to preserve older mechanical assumptions in a world that was rapidly becoming more abstract and observer-driven.

1.2 Photon: A Conceptual Artifact

In modern physics, the photon is treated as an elementary particle—the carrier of the electromagnetic force, the quantum of light. It is described as massless, chargeless, and always traveling at the speed of light. It carries energy proportional to its frequency, as given by Planck's relation:

$E=h\nu$

This definition, widely accepted today, is elegant. It gives a neat mathematical handle on light as energy and allows us to calculate, predict, and model many phenomena. But behind this mathematical clarity lies a deeper conceptual problem—a problem we've never truly addressed.

Einstein's Photon: Newton Repackaged

To understand how the photon came into being, we must look at the historical context. In 1905, Einstein proposed the idea of light quanta—what we now call photons—to explain the photoelectric effect, a phenomenon where light shining on certain metals causes them to emit electrons.

This behavior baffled scientists because it didn't match wave theory. According to wave models, increasing light intensity (i.e., more energy) should eventually eject electrons, regardless of frequency. But experiments showed that:

- Below a threshold frequency, no electrons were emitted, no matter how intense the light.
- Above that threshold, even low-intensity light caused immediate electron ejection.

Einstein resolved this using Planck's idea of quantized energy and Newton's corpuscular intuition. He proposed:

- Light consists of discrete packets—photons—each with energy $E=h\nu$.
- If the energy of one photon is greater than the work function of the material, the photon transfers energy to an electron.
- The electron uses part of that energy to break free; any excess becomes kinetic energy.

It worked. The math added up. And the photoelectric effect was suddenly explainable within the framework of classical conservation laws—energy in equals energy out.

But let's look closely.

What Einstein truly did was wrap Planck's equation around Newton's particle logic. He created "quantized collision logic"—a system where massless particles behaved like billiard balls of energy, striking electrons and disappearing after the impact.

This model was practical, but philosophically shallow.

Where Did the Photon Go?

Here lies the first crack in the model:

If a photon carries energy and momentum but no mass, and it ceases to exist after transferring its energy, then what exactly was it?

- Did it have form?
- Did it occupy space?
- How did it "know" to give all its energy at once?
- Where is the mechanism for this total disappearance?

We do not ask where the photon went. We just say: "It was absorbed." But absorption is not an explanation—it's a mathematical closure. It tells us nothing about the process.

What we're left with is an imaginary particle that:

- Carries energy,
- Travels at light speed,
- Exists only in motion,
- Disappears completely on contact,
- And has zero rest mass, yet measurable momentum.

This is not a particle. It is a story—a convenient myth to maintain the framework of classical causality in a world that clearly demands deeper explanation.

UED Response: The Photon as Disturbance, Not Object

In Universal Energy Dynamics (UED), we take a different approach. We do not deny the effect—but we redefine its origin.

We say:

The photon is not an object. It is a resonance-triggered disturbance pattern—a structured ripple in the universal energy field that arises in response to imbalance.

When a light source vibrates (whether a star or an LED), it creates localized disturbances in the structured vacuum energy. These disturbances propagate not as particles, but as attempts to restore equilibrium.

When that disturbance reaches a material capable of responding—like the surface of a metal—it triggers a reconfiguration of energy within that system. This is what we observe as the ejection of an electron.

No particle arrived. No particle vanished. What occurred was an alignment of energy structure—a resonance condition that made the effect possible.

In this view:

- The photon never travels.
- It does not need to exist independently.
- It is not emitted, but activated.
- Its behavior is not that of a flying object, but of patterned transformation.

Transition: The Photon's Concept Wasn't Wrong—It Was Incomplete

Einstein was not wrong. His model worked within the limits of what science could observe and calculate at the time. But his photon was not the final truth. It was a temporary construct, needed to explain effects that went beyond classical wave theory.

In UED, we retain the predictive power of $E=h\nu$, but we free ourselves from the object-based ontology. We stop imagining photons as massless bullets and start understanding light as dynamic field behavior, structured and revealed only through resonant interaction.

In the next section, we will explore how emission itself is misunderstood, and how realization—not delivery—is the actual mechanism of perception.

1.3 The Difference Between Emission and Realization

The idea that light is *emitted* has long dominated scientific thinking. We picture a glowing bulb, a burning star, or a laser beam—and assume that something is being thrown outward: photons, waves, particles, pulses of energy. We imagine that these things travel through space, reach our eyes, and create the experience we call *light*.

This is the classical view of emission:

A source produces light → the light travels → the observer receives it.

It's simple. It feels intuitive. But it also hides a deeper truth: The observer plays no role in this explanation. The mechanism is one-way, object-based, and devoid of conscious interaction.

But when we shift to the UED framework, the picture becomes more accurate—and far more profound.

Emission vs. Realization: Two Fundamentally Different Models

Let's make the contrast clear:

Classical Emission Model	UED Realization Model
Source emits photons or waves	Source creates a structured energy disturbance
Energy travels through space	Disturbance exists as a potential in the field
Observer passively receives energy	Observer must align with the disturbance
Light exists independent of observation	Light is *realized* only at the point of resonance
Visibility is assumed once light is emitted	Visibility occurs *only when alignment is achieved*

Realization: The Observer-Activated Universe

In UED, we say:

Light is not something sent to you. It is something you unlock through alignment.

When a source vibrates—whether it's a filament, LED, or star—it doesn't shoot particles toward you. Instead, it modulates the energy field around it, creating structured disturbances in the surrounding medium.

But those disturbances become visible—they become *light*—only when your system (eyes, sensors, perception) aligns with that structured vibration.

This is the key:

If there is no resonance, there is no realization.

The "light" may be structurally present, but it does not become perceptible until the observer completes the energetic circuit.

Examples That Prove the Point

Infrared Heaters

- An infrared heater can emit radiation in a room, warming you.
- But you cannot *see* it.
- Your skin resonates with the infrared frequency and feels heat, but your eyes do not align with that part of the spectrum.
- There is energy, but no "light"—until a suitable sensor is brought in.

Ultraviolet Light

- A UV lamp may be active, but you don't see it directly.
- Its effect can be observed in fluorescent materials or sunburns—but only through indirect resonance.
- Again, energy is present—but unless your perception system is tuned, you experience nothing visually.

Quantum Dots

- Quantum dots emit different colors based on their size and structure.
- It's not the particle size itself that "sends" color—it's how

the structure aligns with the energy interaction.
- The emitted wavelength is a result of quantized energy release, structured through resonance—not projectile emission.

Perception Is Not Reception—It's Resonance

What you "see" is not what the universe is sending. It's what your conscious system is tuned to recognize.

This simple insight collapses centuries of assumptions in one breath:

- There is no universal "light" stream moving through space.
- There is only universal energy, structured through vibration, waiting to resonate with intelligent systems.

You are not a receiver. You are a tuner.

You don't see light because it was sent. You see it because you entered its frequency and completed the realization loop.

Closing Thought: Realization as Conscious Completion

This understanding reveals a deeper truth: All that exists around you—light, heat, sound, information—is already vibrating. You experience only what you are capable of resonating with. Everything else remains undetected, unseen, unfelt—not because it isn't there, but because you weren't aligned with it.

In the next section, we explore how this illusion has been built into our everyday understanding of technology, using a familiar example: a simple LED light.

Because even in that glow, we are not witnessing *emission*—We are witnessing the silent realization of structured energy, activated through alignment.

1.4 The Boxed LED and the Myth of Emission

Let's begin with a simple metaphor—a thought experiment so ordinary that we usually ignore its philosophical weight.

Imagine this: You take a brand-new LED bulb, connect it to a battery, and place it inside a sealed, opaque box. No windows, no openings, no sensors. Just the LED, powered and active, glowing quietly in isolation.

Now ask yourself the question:

Is light happening inside the box?

Most would instinctively say, "Of course. The LED is on." But let's look deeper.

If light is something *emitted*, as conventional physics suggests—if photons are being ejected from the LED, flying in all directions at the speed of light—then it would make sense to say, "Yes, light is happening."

But here's the dilemma: No one sees it. No one measures it. No observer is present.

So what's actually happening inside that box?

The LED Doesn't Emit Light—It Restructures the Field

What the LED is doing is not launching particles or shooting out projectiles. It is restructuring the local energy field. The electricity flowing through it causes an internal rearrangement of charge carriers (electrons and holes), and in the process, structured energy disturbances are created in the surrounding field.

But those disturbances don't become *light* until there is a conscious system or tuned detector aligned to realize them.

Without resonance, there is no perception of light. Without tuning, the structure remains invisible.

In this case, the LED is not "emitting light" in the classical sense. It is modulating the vacuum energy structure around it—setting up the *potential* for realization, but not delivering any "thing" to any "receiver."

So again we ask: Is light happening? Or is energy simply structured, waiting for resonance to occur?

Observation = Completion of the Loop

From a UED perspective, light only *happens* when three conditions align:

Source Intent: The LED's internal excitation is the initiating cause—it disturbs equilibrium with structured energy motion.

Field Restructuring: This excitation modulates the surrounding medium—not by ejecting particles, but by shaping the field around it with a distinct vibrational signature.

Observer Tuning: Light becomes real only when an observer (biological or artificial) aligns with this signature and resonates with the structured energy.

Until this loop is complete, light remains a potential—like music waiting to be played by the instrument of your awareness.

What Happens in the Box, Then?

The field is active. The energy is structured. But no realization occurs—because there is no tuning.

It's like a song playing at a frequency no one can hear. The structure exists. The information is present. But the *experience*—what we call *light*—is absent.

So in truth:

Light is not happening inside the box. Energy is being prepared. Light will happen only when resonance occurs—when perception is activated.

The Real Meaning of "Emission"

We've used the word "emission" for centuries to describe how light travels from one place to another. But in the UED framework, this is a linguistic shortcut—a metaphor that became a model, and then hardened into a false law.

Light is not emitted like arrows from a bow. It is realized when a system—source, field, and observer—complete a resonant cycle.

The LED, then, is not a sender. It is a vibrational initiator, capable of triggering realization—but only when an aligned observer joins the field.

Transition to Final Reflection:

This example with the boxed LED exposes the flaw in the particle model of light and illustrates how realization—not emission—is the true mechanism of perception. It also paves the way for a deeper truth: that nothing in the universe "reaches you" unless you align with it. The next time you say "I see the light," you must also recognize:

It is not the light that reached you. It is *you* who entered the light.

In the next section, we'll explore the closing reflection of this chapter: how light is not a message sent from the universe, but a response to your willingness to resonate.

1.5 Closing Reflection

"Light was never the message. You were the receiver waiting to align."

We've now unwrapped one of the most persistent illusions in science and daily life: the belief that light is a thing sent toward us—a package in flight, a signal on its way, an objective stream moving across space. It feels natural to think of light this way.

We've been taught to believe it. We've built technologies around it. But like many intuitive beliefs, it is half a truth made whole by habit.

The deeper truth is this:

Light is not a message the universe sends. It is a moment the universe responds to your resonance.

From Transmission to Realization

All throughout this chapter, we've challenged the notion that light is a *thing in motion*. We've seen how:

- The photon, as a traveling particle, collapses under conceptual scrutiny.
- "Emission," as it's popularly understood, is a projection of classical mechanics.
- "Seeing" is not about catching incoming light—it's about entering alignment with structured energy.

This leads us to the central thesis of the entire book:

Frequency is not something transmitted to you. It is something *realized* through resonance.

Light doesn't "happen" because a source sends it. It happens because you, as the observer, complete the field loop with your intent, perception, or instrumentation.

In other words, you are the one who enters the light—not the other way around.

You Are the Receiver

Every time you "see" light—be it a candle flame, a sunrise, or a blinking LED—it is not because those objects hurled photons in your direction. It is because your system—your eyes, your brain, your awareness—aligned with the structured vibration present in the field.

That alignment is what gives rise to frequency. That alignment is what activates light in your experience. That alignment is the only reason light becomes visible.

This is not metaphor—it is physics, as seen through the lens of Universal Energy Dynamics (UED). And it reframes our relationship with the universe entirely.

The Universe Doesn't Push. It Waits.

There is no cosmic projector throwing out reality. There is only a field of intelligent energy, always structured, always vibrating, always available. What you experience—what you call light, sound, touch, thought—is simply what you are tuned to realize.

This means:

- Nothing is hidden from you.
- But nothing is imposed on you either.
- You see what you are ready to see.
- You receive only what you're capable of aligning with.

And light—one of the most fundamental, mysterious, and misrepresented phenomena in science—is the clearest demonstration of this deeper truth.

Final Words

So the next time you see light, ask not what was sent. Ask what was aligned. Because what you're witnessing is not energy in flight—

It is your consciousness, standing in the right place, at the right frequency, to turn vibration into vision.

Key Takeaways from Chapter 1: The Illusion of Light

Light is not a substance in transit, but a phenomenon of resonance.

What we experience as light is not the travel of photons from a source to an observer, but the *realization* of structured energy when the observer aligns with it.

The photon is a conceptual artifact, not a physical entity.

Born out of the need to explain the photoelectric effect, the photon represents a mathematical convenience that masks the deeper truth of energy field dynamics and observer-dependent resonance.

Emission is a classical illusion; realization is the deeper mechanism.

In the UED framework, light is not emitted from a source and sent to you. It is *activated* only when the observer completes the loop through alignment with structured vibration.

Without resonance, there is no perception—regardless of how much energy is present.

The presence of structured energy is not enough. Without the observer's tuning, no light is seen, no frequency is registered, and no perception is formed.

Observation is not passive reception—it is conscious participation.

You don't merely receive light; you enter it. You co-create the experience by aligning your perception with the vibrational signature already present in the field.

Frequency is not sent—it is awakened.

What we call "frequency" is not an absolute property of light, but a result of resonance. It exists *only* within the framework of time and awareness—only when realization occurs.

Every instance of seeing light is a demonstration of your alignment with the universe.

Light is not proof of something traveling. It is proof that you were tuned to it. What you call light is *your successful resonance* with universal energy.

This chapter redefines our most basic sensory experience—light—not as a delivery from the universe, but as a *dialogue with it.* The next chapter, *The Energy Bubble – Rethinking the Photon*, will explore what light actually is from the inside out, showing how energy seeks balance, not movement, and how the so-called "photon" is better understood as a dynamic field event rather than a particle.

Chapter 2: The Energy Bubble

Rethinking the Photon

"What if the photon was never a thing, but a pattern?"

Imagine this:

A single photon strikes a metal surface. An electron is ejected. The equations are satisfied. Energy is accounted for. And then—the photon vanishes.

Now ask yourself, not mathematically, but philosophically:

If something transfers all its energy and then ceases to exist without a trace—was it ever real in the first place?

This is not a rhetorical game. It's a crack in the foundation of what we call *modern light theory*.

Because if the photon is a particle—then where is its form, its mass, its boundary? If it is a wave—then where is the medium it propagates through? And if it is just a packet of energy—what mechanism allows it to exist, travel, and vanish all without substance?

The photon, as we currently define it, is a model built to solve an effect—not to explain a cause. It was designed to bridge a gap between wave theory and quantum behavior. It does the math. But it does not answer the metaphysical question: What is light *actually*, and how does it *actually* behave?

Energy Transfer Is Real—But the Story Around It May Not Be

This chapter does not deny the reality of energy exchange.

We observe it. We measure it. We see the effects in labs, in photosynthesis, in fiber optics, in the retina of your eye.

But this chapter *does* challenge how we've been conditioned to explain that energy exchange:

- Not as the movement of a particle,
- Not as the collision of a quantized object,
- But as the resonant restructuring of a dynamic energy field.

We don't need the photon as a "thing" to explain energy transfer. We need a new language—one that reflects not what we imagine travels, but what we realize happens when energy seeks equilibrium.

This is where UED reframes the photon: Not as an independent entity flying through empty space, But as a localized pattern—a structured disturbance, Activated by imbalance, shaped by intent, And realized only through alignment with a responsive system.

So let us now take the photon off the pedestal. Not to discard it—but to understand it better by letting go of what it never truly was.

Because what if the photon...was never a thing, but always a pattern of intention in the field?

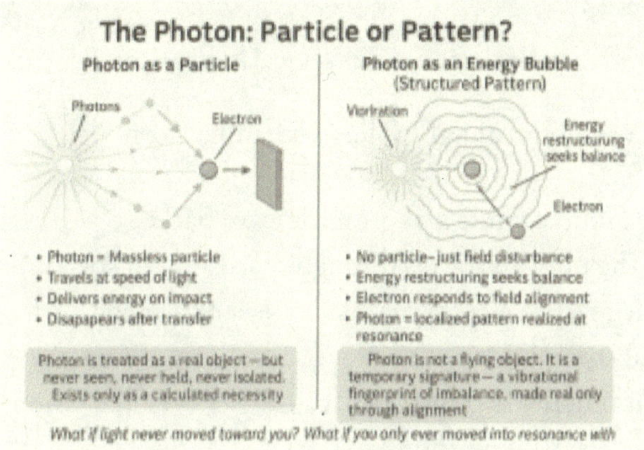

2.1 Structured Energy vs. Particle Logic

To truly rethink the photon, we must first understand why it was ever imagined as a particle in the first place.

The idea didn't emerge out of nowhere—it came from a deeply embedded Newtonian worldview, one that defined the universe as a mechanical system of objects, forces, and motion. In this worldview, everything—from an apple falling to the ground to a cannonball in flight—was explained through particles in motion.

So naturally, when early physicists encountered light and its effects, they reached for the same toolkit.

Light had to be made of something. It had to move. It had to collide. It had to transfer energy.

Hence, the photon was born—not as a directly observed object, but as a conceptual placeholder, rooted in particle logic.

The Mechanical Chain:

Emission → Travel → Absorption

The classical model of light follows a familiar formula:

1. A source emits a photon (or wave packet),
2. The photon travels through space at light speed,
3. It reaches a receiver (like an eye or a metal surface) and is absorbed,
4. Its energy is transferred, causing effects like vision, heating, or electron ejection.

This chain of cause and effect is linear, causal, and mechanical—perfectly suited for Newtonian physics.

But here's the problem: light does not behave like a classical object.

Where the Particle Model Fails

Despite its widespread use, the particle logic applied to photons breaks down under deeper inspection:

1. Directionless Photon Behavior

- Photons do not behave like arrows.
- In quantum experiments (e.g., the double-slit experiment), photons display non-local, wave-like interference, even when sent one at a time.
- They appear to "know" the path options and interfere with themselves, suggesting they are not *localized particles* at all.

2. Instantaneous Effects (e.g., Tunneling)

- In phenomena like quantum tunneling, photons can appear on the other side of a barrier faster than light would allow.
- This implies there's no particle crossing space—but rather, field restructuring occurring through a different mechanism.

3. Momentum Without Mass

- Photons are said to carry momentum (p=E/c)—but have no rest mass.
- Momentum, classically, is a product of mass and velocity.
- If the photon has no mass, what exactly is carrying that momentum?
- This contradiction is rarely explained, only accepted.

These inconsistencies show that the particle model of the photon is a mathematical workaround, not a reflection of what's really happening in the field.

Enter UED: Energy as Structured Potential

In Universal Energy Dynamics (UED), we take a fundamentally different approach.

Energy is not a substance. It is a condition. It does not travel. It restructures.

Here's how UED reframes the idea:

1. Energy Exists as Field Potential

- The universe is not empty—it is a structured, dynamic medium of stable energy.
- Every event is not caused by an object entering space—but by the restructuring of local energy patterns.

2. Structure Defines Behavior, Not Form

- What we observe as "photon effects" are actually field responses to structured vibration.
- Structure—not a particle—determines:
    - How energy behaves
    - Where it interacts
    - What effect it produces upon resonance

3. Patterns Appear Quantized at Resonance

- Energy doesn't *move* from place to place.
- Instead, structured energy resonates across space, and appears quantized (like a photon) only when resonance occurs between source and receiver.
- The quantization is not a property of a particle—it is a perceptual effect of resonance timing and tuning.

The Takeaway

The photon isn't a flying object. It's a pattern of response within a field, made visible only when resonance completes the loop between:

- Source intent (energy disturbance),
- Field structure (medium behavior),

- Observer alignment (resonant decoding).

There's no need to imagine something flying, hitting, and disappearing. There's only energy trying to balance itself—and you happen to witness it when you're tuned in.

In the next section, we'll show how no actual transfer or collision is required to explain this behavior—and how light becomes visible not through delivery, but through localized restructuring in the medium itself.

2.2 No Collision, No Transfer—Only Restructuring

Let's now revisit the photoelectric effect, the very experiment that gave rise to the photon's conceptual identity. It was this phenomenon—electrons being ejected from metal surfaces when light shines upon them—that Einstein used to propose the quantum of light: the photon, carrying energy $E=h\nu$.

But beneath the surface of this neat explanation lies a deeper truth that's rarely addressed:

There is no observable impact. No visible collision. No actual object transfer. Just a result. Just a response. Just a restructuring.

What Really Happens in the Photoelectric Effect?

Here's what we observe in the lab:

- A clean metal surface is placed inside a vacuum tube.
- Light of a certain frequency (not intensity) is directed at it.
- If the frequency is below a certain threshold, no electrons are emitted, no matter how bright the light.

- If the frequency meets or exceeds that threshold, electrons are ejected immediately, even with low-intensity light.

The classical (pre-Einstein) wave model couldn't explain this. If light were purely a wave, increasing intensity should eventually provide enough energy to eject electrons—even at lower frequencies. But that didn't happen.

So *Einstein proposed a new model*: Light arrives in packets (quanta), and if a single photon has enough energy, it transfers that energy to a single electron in a one-to-one impact, ejecting it instantly.

This model worked mathematically. But conceptually, it reintroduced a Newtonian mindset—the idea that energy is a projectile, a bullet, a particle.

But now, with UED in hand, we must ask:

Why must we imagine a particle at all? Why not explain what we see in terms of field behavior?

UED Reframing: Ejection Is Not an Impact—It's a Response

In Universal Energy Dynamics (UED), we do not deny the observed effect. The electron is ejected. The current does flow. The timing is immediate.

But the explanation is very different:

- There is no photon flying through space.
- There is no collision at the surface.
- There is no energy packet being absorbed like a courier package.

Instead:

- The light source vibrates in structured energy.
- That vibration modulates the surrounding field, sending out a resonant pattern.
- When that pattern aligns with the internal structure of the material surface (at the correct frequency), a localized field reconfiguration occurs.
- This restructuring releases an electron, which is already a part of the surface's energy system.
- No transfer. No impact. Just triggered release through resonance.

There Is No Delivery—Only Activation

The idea of a photon "carrying" energy from source to electron implies a transport mechanism—an object-like logic rooted in classical physics. But what if the energy wasn't carried, but already present in the field—waiting for the right condition to activate it?

This is what UED proposes:

The ejected electron is not the result of something arriving. It is the result of something aligning.

There is no need to imagine the photon "disappearing" after giving its energy. There was no object to disappear. There was only a vibrational signature, a field resonance, and a reconfiguration of structure.

"Absorption" Is a Misleading Word

The language of modern physics is still deeply mechanical—even in the quantum realm. We say things like:

- "The electron absorbed the photon."
- "The photon was destroyed."
- "The energy was transferred."

But these are narrative shortcuts—anthropomorphic terms that mask what's actually happening.

Nothing is absorbed. Nothing is destroyed. Nothing is handed over like currency.

Instead: The system reorganizes itself in response to resonant disturbance. The field adjusts. The structure shifts. The electron reconfigures.

It is not a mechanical exchange. It is a field-level event, born not out of collision, but out of coherence.

Final Thought

So the photoelectric effect doesn't demand a photon. It demands a frequency match. And that match doesn't cause an impact. It causes release.

What we've called "light transfer" is nothing more than the activation of a latent possibility within the structured field.

In the next section, we will explore this idea more deeply by turning our attention to the vacuum itself—not as empty space, but as a dynamic medium that carries structure, vibration, and intent—even when we believe nothing is there.

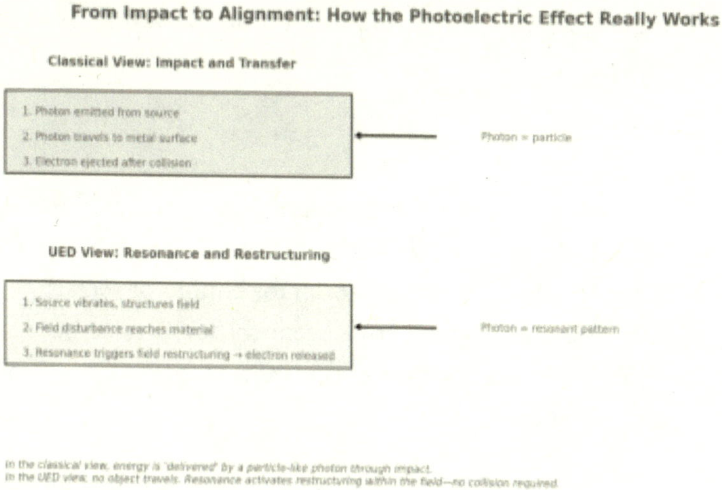

2.3 The Vacuum as Medium, Not Empty Space

At the heart of nearly every modern physics equation is an invisible assumption—so embedded that we rarely question it:

That space, unless filled with matter or energy, is empty.

We call it vacuum. We treat it like zero. We build equations around it as though it is absence—a neutral background in which particles fly and waves propagate.

But this assumption is not only philosophically flawed—it is physically limiting.

The Myth of Empty Space

The classical and quantum models often present vacuum as a backdrop—a neutral, passive void. Light, in this model, is said to move through space at 299,792,458 m/s because "nothing" is in its way.

But wait.

- How can a wave travel in a vacuum if there's no medium?
- What is the wave oscillating?
- What "carries" a massless photon for billions of years?
- Why does the vacuum itself seem to fluctuate, as seen in phenomena like Casimir effect, vacuum polarization, and zero-point energy?

Clearly, space is not empty. It is active, structured, and full of hidden dynamics.

Vacuum Is Structured, Stable Energy

In the UED framework, we redefine vacuum as follows:

Vacuum is not nothingness. It is the ground state of structured energy.

It is not absence—it is balance. A vacuum is not void of action—it is saturated with intelligent symmetry that holds all potential for interaction.

From this perspective:

- Space is a medium.
- Its baseline state is not static—it is a perfectly poised equilibrium of energy, capable of responding to disturbance.
- Every interaction we observe—light, heat, gravity, fields—is an expression of temporary restructuring within this dynamic medium.

Waves Without Particles: Propagation Through Structure

In everyday life, sound travels through air. You don't need to imagine sound particles—what moves is a disturbance through a structured medium.

Now apply that analogy to light:

- The "photon" does not travel through empty space.
- Instead, the field is disturbed by a vibrating source.
- That disturbance resonates outward through the structured vacuum, like ripples on a lake.
- No object moves. Only patterns shift, seeking to restore equilibrium.

Just as sound requires air, light requires structure in the vacuum. This structure doesn't need to be visible. It only needs to be intelligently connected—capable of transmitting resonant patterns across spacetime without mass or substance.

Photon "Travel" Is a Field Ripple, Not Object Motion

When we say a photon moves from a distant galaxy to your telescope, we imagine it like a bullet making a straight-line journey through space.

But in UED terms:

- The source vibrates and modulates the vacuum structure.
- That modulation travels not as a thing—but as a field ripple.
- When your system (the telescope, your retina, a sensor) is tuned to resonate with the incoming pattern, realization occurs.

- Light is not *delivered.* It is unlocked at the point of resonance.

There was no photon flying for billions of years. There was only the field, shaped by a source, holding a structured memory of disturbance, awaiting the right condition to be realized again.

Final Thought for This Section

We don't need to explain light as particle motion through emptiness. We only need to recognize that:

The vacuum is a field of potential. What we see as "travel" is actually a ripple in this field, seeking to return to balance.

Light doesn't move across space. Space restructures itself, and you observe that restructuring only when your system resonates with it.

In the next section, we will tie all of this together with a new definition of light—not as energy in transit, but as a disturbance seeking balance. That is what the photon really is: a field intention, not a flying object.

2.4 Light as Disturbance Seeking Balance

By now, we've moved well beyond the idea of light as a traveling particle. We've dismantled the notion of a flying photon, redefined vacuum as a dynamic structure, and reframed photoelectric events as resonant field responses, not collisions.

Now we arrive at the heart of the matter:

Light is not an object in motion. It is a disturbance in the field—a structured vibration—seeking to restore balance.

This is not just poetic reinterpretation. It is a physically grounded, scientifically essential reframing. Let's walk through this carefully.

Light as a Structured Vibration

When a source (like the Sun, an LED, or an atom) becomes energized, it doesn't "emit" a packet. It experiences a shift in its energy structure, creating a disturbance—a vibrational ripple—in the surrounding field.

This ripple is not random. It is structured, coherent, and patterned based on:

- The energy level involved
- The nature of the source
- The field dynamics of the medium (which we now know is never truly empty)

That structured ripple is what we observe as light.

And like all disturbances in a stable system, its core drive is simple:

To return the system to equilibrium.

Not Motion, but Resonant Adjustment

This is the critical shift in understanding: Light does not move from A to B like a messenger. Instead, it propagates as a systemic field reaction—a real-time restructuring, much like pressure waves in air.

- There is no photon flying from source to destination.
- There is no "arrival" of a thing.
- There is only resonance propagating outward, and restoration of balance as the system reconfigures itself.

This explains why light:

- Appears to travel at a constant speed
- Behaves with perfect directional consistency
- And yet carries no mass or form

It behaves as if a particle is involved—but only because the pattern is so sharply defined, so discretely realized, that it *appears* quantized to the observer.

But in truth:

What we call a photon is just a signature—a point of field restoration that we happen to be tuned in to witness.

What This Explains Better Than Particle Theory

Let's address the core behaviors that become instantly clearer in this model:

1. The Apparent Speed of Light

In classical physics, the speed of light (ccc) is a constant—independent of source motion or observer speed. This has puzzled scientists for generations.

But in UED, this speed is not the velocity of a flying particle.

It is the natural rate at which structured disturbances propagate through the vacuum field—a property of the medium itself, like the speed of sound in air.

Light's speed is constant not because particles are miraculous—but because the field is uniform in its ability to rebalance structured energy.

2. Directionality Without Projectiles

Light behaves with precision: it reflects, refracts, diffracts, focuses, bends.

These behaviors are often credited to photons being directional. But that's just an interpretation.

In the UED model, directionality emerges from how structured disturbances ripple through the field, guided by:

- Boundary conditions
- Field gradients
- Resonant structures within materials

The "path" of light is not a photon's journey. It is a field ripple tracing the path of least energetic resistance—just like water around a rock or heat through a gradient.

3. Energy Release Without a Carrier

One of the deepest mysteries in conventional physics is how massless photons carry energy and momentum.

UED's answer: they don't.

The energy isn't *carried.* It is localized and released as the field restores itself.

In a photoelectric effect, for example:

- The material doesn't "receive" a photon.
- It responds to a resonant vibration pattern.
- The energy release (e.g., the ejected electron) is a field-based restructuring, not a particle-particle interaction.

There is no carrier. There is only consequence.

Final Reframing:

Light is not a courier—it is a correction. Not a projectile—but a pulse of balance, emerging from disturbance.

We do not see a "thing" arriving. We witness the field completing a cycle of coherence—a feedback event between source, medium, and observer.

In the next chapter, we will now take the bold step of redefining the photon entirely—not as a particle or wave, but as a resonant signature—a realization at the intersection of vibration and perception.

Because what we've called a photon...was never a particle. It was a pattern of realignment in a field that never stopped vibrating.

2.5 Transition to Chapter 3 – Photon Replaced: The Resonant Signature

We began this chapter with a bold question: What if the photon was never a thing, but a pattern?

And throughout this journey, we have steadily dismantled the need for the photon as a traveling particle, a massless projectile, or a carrier of energy. We have seen how the observable effects traditionally attributed to photons—electron ejection, energy release, light behavior—can be more elegantly and accurately explained by a different mechanism altogether:

Structured energy disturbances in a dynamic field, seeking balance through resonance.

In this view, nothing moves from source to destination. There is no packet being hurled through empty space. There is no need for a carrier, a collision, or an absorption event.

So the natural question now becomes:

If There's No Particle, and No Emission, Then What Gives Light Its Identity?

- What gives us the consistent, measurable behavior of light?
- What explains its apparent discreteness, directionality, and predictability?
- If there's no "photon," how do we account for the regularity of effects that still appear quantized, precise, and reproducible?

These are not trivial concerns. They are the core of the scientific mindset: we don't just want to observe—we want to explain. We want to identify *what* light is, *why* it behaves the way it does, and *how* we can meaningfully work with it.

And this brings us to the next leap in understanding:

What if light is not defined by what it "is," but by what it "does" when a system resonates? What if light is not an entity—but a resonant signature, realized in the field, decoded by the observer?

From Particle to Pattern, From Object to Outcome

This is where Chapter 3 will begin: not with another theory of "what the photon is made of," but with a more powerful idea—

That light's identity is not derived from a form, but from a field interaction that follows consistent laws of resonance, structure, and equilibrium.

In short:

The photon is not a particle. It is the signature of a successful alignment.

And it is time we start treating it that way.

Key Takeaways from Chapter 2: The Energy Bubble – Rethinking the Photon

The photon is not a fundamental object, but a conceptual artifact.

Originally introduced to explain the photoelectric effect within a classical framework, the photon model is based on particle logic inherited from Newtonian mechanics—not on direct observation or universal necessity.

Classical particle-based logic breaks down when applied to light.

Phenomena like photon momentum without mass, quantum tunneling, and interference patterns reveal that light doesn't behave like discrete particles in motion.

UED redefines energy as structured potential, not as moving entities.

Energy exists in the field as dynamic structure. What appears quantized is actually the result of localized resonance—not the delivery of "energy packets."

The photoelectric effect does not require impact or transfer.

Electron ejection is not caused by a collision with a flying photon, but by **field restructuring** triggered by resonance at a specific frequency.

Vacuum is not empty space—it is a dynamic, structured medium.

What we call "vacuum" is a state of perfectly balanced, stable energy. Light propagates not through emptiness, but through structured field modulation—similar to how sound ripples through air.

Photon behavior is better explained as disturbance seeking balance.

Light is not in motion. It is a **systemic ripple** caused by energetic imbalance and resolved through resonance. Its apparent speed and directionality are features of field response, not object trajectory.

There is no transfer of energy—only realization through alignment.

The effects we attribute to photon impact are actually field-level responses. Energy isn't handed over. It is **unlocked** when the observer or system matches the vibration of the source.

Light's identity comes not from being a thing—but from being a signature.

What we perceive as a photon is not an object in space, but a **structured resonance pattern** in the field—realized only through a triadic relationship between source, medium, and observer.

This chapter redefines the photon from a hypothetical particle to a real-time vibrational event, and reframes light not as energy in motion, but as a signature of structural imbalance resolved through resonance.

In Chapter 3, we'll go further—not just dismantling the photon model, but replacing it with something more profound:

The Resonant Signature—light as the universal echo of alignment.

Chapter 3: The Photoelectric Effect

A Misunderstood Resonance

"The Experiment That Defined a Century"

If there was one experiment that pushed classical physics into the quantum realm, it was the photoelectric effect.

Simple in setup, profound in implication, this experiment shattered the comfortable assumptions of wave-based light theory and demanded a new way to explain how light and matter interact.

What Was the Photoelectric Effect?

At its core, the experiment involved shining light onto a metallic surface in a vacuum and observing whether electrons were ejected as a result.

What was observed defied classical predictions:

- Increasing the intensity (brightness) of light did not necessarily eject electrons.
- Only when the frequency of light crossed a certain threshold did electrons begin to be emitted—regardless of the intensity.
- Below that threshold, no amount of light could make it happen.

This broke with the expectations of classical wave theory, which said:

"The more energy (intensity) you apply, the greater the effect."

But reality didn't agree.

Einstein's Revolutionary Explanation

In 1905, Albert Einstein offered a radical interpretation. He proposed that light was not just a wave, but came in discrete, quantized packets of energy—what we now call photons.

Using Planck's relation $E=h\nu$, Einstein argued that:

- Each photon has energy proportional to its frequency.
- If that energy is equal to or greater than the work function of the material (i.e., the minimum energy needed to release an electron), then an electron would be ejected.
- Otherwise, nothing happens—no matter how many photons arrive.

This was a bold departure from classical reasoning—and it worked.

Einstein's explanation not only matched the experimental data perfectly but also helped launch quantum theory into serious scientific discourse.

In 1921, he was awarded the Nobel Prize in Physics—not for relativity, but for this very explanation of the photoelectric effect.

Why This Chapter Matters

Let's be clear: Einstein wasn't wrong. The results were correct. The mathematics worked. The explanation was consistent within the framework available at the time.

But like all scientific models built under constraints, it came with assumptions—assumptions we now know can and should be challenged.

Einstein's photon was a brilliant conceptual solution, but it was still:

- Based on particle logic,
- Framed by Newtonian intuition,
- And rooted in the idea of energy transfer via collision.

The Purpose of This Chapter

In this chapter, we do not discard the photoelectric effect. We honor its importance and preserve the value of its outcome.

But we also re-examine its cause through the lens of Universal Energy Dynamics (UED):

- We challenge the idea of "photons" being real, moving particles.
- We reinterpret the electron's ejection not as a mechanical impact—but as a resonant release within a structured field.
- We shift the conversation from energy packets traveling to vibrational patterns aligning.

This isn't just a reinterpretation. It's a clarification of the mechanism—one that better reflects what the universe actually seems to be doing.

Because maybe... just maybe...

The photoelectric effect never needed a particle at all. It only needed a match. A frequency. A moment of resonance.

3.1 Einstein's Pragmatic Patch

In the early 20th century, physics stood at a threshold. Classical theories, long considered complete, began to show signs of limitation. The photoelectric effect was one such crack in the edifice—a stubborn observation that refused to conform to wave-based models of light.

At this moment of conceptual tension, Einstein stepped in—not with philosophical revolution, but with pragmatic necessity.

The Planck Foundation

Einstein's explanation of the photoelectric effect did not emerge in isolation. It leaned heavily on the earlier work of Max Planck, who had proposed the quantization of energy to explain blackbody radiation—another anomaly classical physics couldn't handle.

Planck had introduced the radical idea that energy was not continuous, but instead came in discrete units or "quanta", where:

$E=h\nu$

- E is the energy of the quantum
- h is Planck's constant
- ν is the frequency of the radiation

Einstein borrowed this formula but extended its implications far beyond Planck's conservative stance. He proposed that light itself—not just energy within matter—was quantized.

The Photon Enters the Scene

Einstein declared that light behaved as though it were made up of individual particles, each carrying a quantum of energy determined by its frequency.

These particles—soon to be named photons—could strike electrons on a metal surface. If the energy of the photon exceeded the work function (the minimum energy needed to release an electron from the metal), the electron would be ejected.

If not, no emission occurred, regardless of the number of photons or the light's intensity.

This proposal explained:

- The threshold frequency required for electron ejection
- The instantaneous nature of the effect (no time delay)
- The independence from light intensity

The math worked. The predictions matched the data. Problem solved—at least on paper.

But Was It an Explanation—Or a Fix?

Einstein's solution, while brilliant, was pragmatic—not philosophical. It wasn't a deep rethinking of energy or light. It was a mathematical patch that fit the experimental outcome into the existing structure of cause-effect physics.

He took:

- Planck's energy quantization
- Newton's particle logic
- And classical mechanics' thirst for causal events

...and stitched them together into a convenient conceptual narrative:

Light is made of particles. Particles carry energy. Energy is transferred on contact. Electrons are released.

It was elegant and useful—but not necessarily true.

Einstein himself was cautious. He knew this was a model—not a fact. But over time, the model hardened into a belief.

The photon, once a helpful metaphor, became a declared entity.

The UED Critique: Photon as Conceptual Stopgap

From the perspective of Universal Energy Dynamics (UED), the photon was not a discovery. It was a translation tool—a bridge between a new observation and an old worldview.

We critique it not because it was wrong, but because it was incomplete.

The photon is not a particle of light. It is the name we gave to an outcome we didn't yet understand.

In UED:

- Light is not made of things. It is made of structured energy disturbances.
- What we call a photon is the realized resonance between source vibration and field structure.
- There is no transfer. No travel. No impact.
- There is only alignment, and the resulting reconfiguration of field potential.

Einstein's photon helped physics move forward—but it also trapped it inside a particle-bound mindset that is now overdue for release.

Final Thought:

Einstein's work was undeniably revolutionary. But it was also bounded by the logic of its era—an era where everything had to move, collide, or be delivered.

UED offers a broader lens:

We don't need particles to explain energy exchange. We only need structure, resonance, and the field's inherent drive to balance itself.

And with this new lens, we are ready to revisit the most important detail Einstein revealed—frequency—and ask:

Who really sets it? And why does it matter more than intensity?

3.2 Threshold Frequency: Who Sets It?

One of the most compelling aspects of the photoelectric effect—and the most confusing to classical physicists—was not that electrons were ejected when light hit a surface, but how selectively it happened.

More light didn't always mean more effect. Brighter light didn't guarantee results. It wasn't about how much light—but what kind of light.

The Classical Paradox: More Energy, No Effect

According to classical wave theory, increasing the intensity of light should increase the energy delivered to the surface.

- Shine a brighter light, deliver more energy per second.
- Eventually, the electrons should absorb enough energy and be ejected—even if the frequency is low.
- Like heating water: more heat = faster boiling.

But the experiment said otherwise.

- No matter how intense the light, if its frequency was below a certain point, no electrons were emitted.

- Even dim light—if above that threshold frequency—could cause instantaneous ejection.

This defied logic. It implied that something other than sheer energy quantity was governing the interaction.

So physics asked:

Why does frequency matter more than intensity?

The Photon Explanation: High-Frequency = High-Energy

Einstein's photon model solved the paradox by introducing quantized energy packets:

$E=h\nu$

- The higher the frequency (ν\nuν), the more energy each photon carries.
- If a photon's energy is below the metal's work function, no electron escapes—no matter how many photons you throw.
- If a single photon's energy exceeds the threshold, it can release an electron—instantly.

This made the math work. But it also forced us to accept:

- That each photon is an autonomous energy carrier.
- That the threshold is set by the material, but the key is frequency—not the number of photons.

Still, a deeper question remains unanswered:

Why does only a specific frequency allow the interaction to occur? Why can't energy just accumulate gradually and trigger a release?

This is where UED offers a cleaner answer.

UED View: Threshold Frequency Is Not Intrinsic to Light

In Universal Energy Dynamics, frequency is not a trait of a traveling particle—it is the signature of a vibrational structure in the field.

Here's the key distinction:

- Classical view: Light carries energy to the surface.
- UED view: Light is a patterned disturbance, and the surface responds only when that pattern matches its internal structure.

The threshold frequency is not set by the light source. It is set by the resonant potential of the material's internal energy configuration.

It's like a lock-and-key relationship:

- The lock (material) will only respond to a specific frequency shape (the key).
- Not because of how strong the key is—but because of how well it fits.

Frequency as Matchmaker, Not Hammer

In classical physics, energy is like a hammer: the more force, the more effect.

But in UED, energy is more like a tuning fork. You don't need to hit harder—you need to hit the right note.

Frequency is not a measure of how powerful light is—it's a measure of how well it aligns.

The surface doesn't respond because it's been hit hard enough. It responds because the vibrational field of the incoming energy has synchronized with its own latent structure.

Only then does the system restructure itself and release an electron—not due to impact, but through resonance-triggered reconfiguration.

Reframing the Threshold: Alignment, Not Transfer

The entire idea of a "threshold" in UED is not about minimum energy required—it is about the frequency window within which alignment becomes possible.

This explains:

- Why higher frequency light works regardless of intensity: it resonates.
- Why lower frequency light fails regardless of brightness: it doesn't align.
- Why electron ejection is instant: resonance is immediate—not cumulative.

Final Thought

So when we ask, "Who sets the threshold frequency?" The answer is: the material does—but not based on mechanical resistance or mass-energy ratios.

It sets the threshold based on its internal resonant structure—its unique energy architecture.

And light doesn't "overcome" this structure. It simply finds harmony with it—or it doesn't.

In the next section, we'll go even deeper into this harmony by showing that material doesn't react like a passive object being struck—it participates in the realization through active alignment.

3.3 Material Doesn't React—It Aligns

Physics, for centuries, has explained phenomena through a familiar mechanical lens: Object A applies force → Object B reacts.

This paradigm works well for billiard balls, levers, and classical collisions. But when applied to quantum-scale interactions—especially something as subtle as the photoelectric effect—it becomes misleading.

Classical Assumption: Impact = Reaction

In the traditional explanation of the photoelectric effect, light is imagined as a stream of particles (photons), each carrying energy. When a photon strikes a metal surface:

1. It collides with an electron,
2. Transfers its energy in a single hit,
3. If the energy is enough, the electron is ejected.

This logic implies:

- A material is passive, waiting to be hit.
- Light behaves like a hammer, and the electron is a nail.
- Energy moves like a projectile, with cause and effect tied to physical contact.

But this model is based on metaphor—not mechanism.

It forces a collision narrative onto a phenomenon that behaves more like vibration, music, and resonance than mechanics.

UED View: Not Reaction—But Alignment

In Universal Energy Dynamics, the material is not struck—it is stimulated.

It doesn't react like a door being kicked open. It responds like a tuning fork vibrating when the right note is played.

The photoelectric effect is not about force—it's about fit.

What matters is not how much energy is thrown at the surface, but whether the incoming structured field pattern aligns with the internal vibrational spectrum of the material.

How Alignment Works in UED

The photoelectric effect happens only when two criteria are met:

1. Field Resonance

- The source (e.g., light-emitting body) must generate a structured vibration in the field.
- This structured energy must match the resonance band of the surface material.
- This match is not about energy quantity—but about pattern compatibility.

2. Configuration Enables Restructuring

- The material's internal structure—its atoms, electrons, and field interactions—must be capable of reorganizing under the influence of this resonant vibration.

- When matched, the system undergoes a restructuring, and the electron is released.
- Not "hit" out. Not "knocked" loose.
- But realigned and liberated through coherent energetic reconfiguration.

The event is simultaneous, non-destructive, and non-impact-driven.

Electron Ejection: Not Collision, but Completion

In this light, the ejected electron is not a victim of impact—it is the consequence of a successful alignment.

Think of it like this:

- A locked mechanism is full of tension.
- You don't break it open. You turn the key.
- The door swings open—not because you forced it—but because you understood its structure.

That is what light does—not deliver force, but offer structure that unlocks hidden potential.

The electron is released not from being hit—but from being harmonized out of the system.

A Paradigm Shift in Action

This shift—from reaction to alignment—reorients how we think about all field-based interactions:

Classical View	UED View
Energy is transferred by collision	Energy is realized through resonance
Material is passive and reactive	Material is structurally participative
Emission is caused by impact	Emission is triggered by pattern match
Frequency is energy per packet	Frequency is a resonance condition

This also explains why:

- No time delay occurs: alignment is instantaneous.
- No cumulative effect occurs: misalignment can't build up to success.
- No partial ejection occurs: it's all or nothing—resonance or silence.

Final Thought

The photoelectric effect doesn't happen because something *lands*. It happens because something fits.

And that fit is not mechanical—it is structural. Not explosive—but elegant.

The electron is not ejected by an impact—it is released by a realization. A moment when structure, vibration, and alignment converge.

In the next section, we will explore how even our measurements in these experiments are shaped by this principle—revealing not objective transfer, but subjective induction within resonant systems.

3.4 What We Measure Is What We Induced

One of the most powerful illusions in science is this:

That measurement reveals reality.

We build instruments, run experiments, collect data, and assume that the numbers we see on our screens are direct reflections of what the universe *is* doing.

But what if they're not?

What if what we're measuring is not a complete picture of what occurred, but a localized resonance—a manifestation of our system's capacity to respond?

In the case of the photoelectric effect, this distinction becomes critical.

Instruments and Observers: Tuned Systems, Not Neutral Windows

Whether it's the detector in a lab, a sensor on a satellite, or the retina in your eye—every perceiving mechanism is a tuned structure.

- Instruments detect within specific frequency bands.
- Human senses operate within biologically constrained response ranges.
- Even mathematical models are based on theoretical tuning—filters of thought based on existing frameworks.

This means we never observe "everything". We only observe what our tools and systems are structured to resonate with.

What We Call Kinetic Energy Is a Resonance Artifact

In the photoelectric effect, when light of the right frequency hits a material and electrons are ejected, we observe these electrons having specific kinetic energies.

Classical thinking says:

"The photon hit the electron and transferred energy, which now shows up as kinetic motion."

But UED asks:

"Why assume transfer? Why not consider that the system restructured, and what we measured was the residual energy of a successful resonance event?"

In this light, kinetic energy is not a leftover from impact. It is a trace signature—the final, visible artifact of field realignment.

The material didn't receive a push. It underwent a rebalancing, and the ejected electron is the part of that system which now behaves independently.

We didn't measure a transfer—we witnessed a consequence of alignment.

Measurement Reflects System Capacity, Not Universal Truth

This is one of UED's foundational claims:

Measurement does not tell us what happened in the universe. It tells us what our system was capable of realizing.

Every measurement is a snapshot of realization, taken within:

- The boundary conditions of the apparatus
- The limitations of the observer
- The assumptions of the model

This means our instruments are not cameras—they are participants in the process. They don't passively record—they resonate with available structure and reveal what that resonance allows to appear.

So when we "see" the photoelectric effect:

- We are not witnessing energy flying and hitting.
- We are witnessing a field response, made visible through engineered resonance.

The Photoelectric Setup: An Alignment Game

From beginning to end, the photoelectric effect is not a test of delivery—it is a test of alignment:

Component	Classical View	UED View
Light source	Emits photons	Vibrates structure into field
Metal surface	Receives energy via impact	Responds to matching resonance
Electron	Absorbs energy and escapes	Released due to structural re-tuning
Detector/Measurement	Captures evidence of energy transfer	Reveals visible trace of alignment

There is no courier, no transaction, no impact. There is field pattern → structural match → localized realization.

And what we *record* is not the universe's objective output—It's what we were able to induce, given our system's architecture.

Final Insight

The real innovation of the photoelectric effect wasn't the photon. It was the discovery that reality responds to structure—not force.

We do not measure energy transfer. We measure energetic resonance—when field, matter, and awareness complete a triad of alignment.

What Next

As we move forward, we must now explore the consequences of this realization:

- If frequency governs resonance,
- If alignment—not force—triggers interaction,
- And if measurement is a tuned realization...

Then we must ask:

What is energy, really? Is it a quantity to be moved—or a condition to be unlocked?

Next chapter begins with this question and expands our understanding of energy itself—not as substance, but as a vibrational expression of intent.

3.5 Rethinking Threshold and Energy

The photoelectric effect has long been hailed as a victory of quantum mechanics over classical physics. And indeed, it was—but not for the reasons we often assume.

It didn't prove that light is made of particles. It didn't confirm that electrons are "hit" out of matter. What it really revealed—quietly, indirectly—was that interaction is governed by alignment, not aggression.

And now, having deconstructed the idea of impact, transfer, and force, we arrive at a deeper and more fundamental question:

If Threshold Frequency Is Not an Absolute, Then What Is Energy—Really?

What we saw in the photoelectric effect was not a hard limit on nature. It was a boundary condition—a relational window of resonance compatibility between an incoming field disturbance and a receptive structure.

This means:

- The threshold was not a property of the light.
- Nor was it a universal constant.
- It was a manifestation of structural alignment between the source and the observer (material).

So now we must ask:

If frequency defines resonance, and resonance defines outcome, then isn't energy itself—as we know it—just a reflection of how well two systems align?

What we call "energy transfer" may actually be energy expression—the realization of potential when structures find coherence.

This redefinition turns energy from a thing into a process. From a quantity to a quality. From a force in motion to a pattern in harmony.

From Quantity to Structure: Energy as Intent Made Form

In traditional physics, energy is defined in scalar terms:

- Joules, electron volts, kilowatt-hours.
- It's a "stuff" to be stored, moved, consumed.

But in UED, energy is not a substance. It is a condition—a structured vibration that expresses itself only when certain intentional patterns align.

It doesn't need to be delivered. It doesn't need to be transferred. It needs to be activated through compatibility—just like music needs instruments, not more volume.

This leads to a revolutionary shift in thinking:

Energy is not what moves. Energy is what emerges when structure meets intent.

A New Cosmic Language

If we accept this, then everything changes:

- Light is not a carrier—it is a conductor of resonance.
- Matter is not a recipient—it is a participant.
- Energy is not a fuel—it is a signature of alignment in the universal field.

And if this is true, then the entire cosmos—from photon interactions to gravitational forces—must be reinterpreted not as collisions between particles, but as expressions of alignment across dynamic structures of intent.

The next chapter takes us fully into this vision. We leave behind the language of force and impact and embrace a more precise model—one where the photon is no longer needed, because we finally understand that:

The universe does not throw energy—it sings. And only those who are tuned can hear.

In next chapter, we will formally replace the concept of the photon with what it always truly was: A resonant signature. Not an object. Not a packet. But a pattern of realized potential—a cosmic echo of structural coherence.

Key Takeaways from Chapter 3: The Photoelectric Effect – A Misunderstood Resonance

Einstein's photon was a brilliant but pragmatic patch.

While his explanation matched experimental results, it inserted a hypothetical particle into the framework to preserve classical cause-effect logic—without resolving the deeper question of *why* frequency mattered more than intensity.

The threshold frequency is not intrinsic to light—it's relational.

The effect doesn't depend on how much energy is thrown, but on whether the light's structured vibration aligns with the material's internal resonance band.

Material surfaces don't react—they align.

The photoelectric effect is not caused by energy being transferred to electrons via impact. It is a field-level reconfiguration, triggered by coherence between the incoming pattern and the material's latent vibrational architecture.

Electron emission is the result of internal restructuring, not external force.

The electron is not "knocked out," but liberated through a matched vibrational pattern that allows the material system to shift its state.

What we measure is what our system is tuned to detect.

Instruments and perceptions are not neutral—they are resonant systems. What we call kinetic energy or photoelectric current is not proof of particle behavior, but of induced structural alignment made observable through our tools.

The entire experiment is an alignment mechanism, not a delivery test.

There is no courier particle delivering energy to matter. There is only structured energy seeking balance, and a surface that either aligns—or doesn't.

Energy is not transferred. It is realized.

What we measure is not what was sent. It is what became visible when intent, vibration, and structure synchronized.

This resonance model sets the stage to redefine the photon—not as a particle, but as a pattern.

Chapter 4 will show how the photon is better understood as a resonant signature, revealing the true nature of energy as structured potential, not moving massless objects.

Chapter 4: Frequency Emerges with Observation

"You don't observe frequency. You define it."

Take a moment and ask yourself: What is a frequency?

Most would say: "It's how often something happens in a given time." In physics, we measure it in hertz—cycles per second. A sound wave at 440 Hz? That means 440 cycles in one second. A gamma ray at 10^{20} Hz? Trillions of oscillations per tick of the cosmic clock.

But stop and consider the quiet assumption buried beneath all this:

What is a "second"—if there is no observer to count it?

Who determines the beginning and the end of a cycle? Who chooses the time window in which these cycles are counted? What makes a vibration become a frequency, and not just a silent oscillation lost in the fabric of reality?

The Hidden Role of the Observer

Here lies the paradox:

You think you're observing frequency—but in truth, you're the one constructing it.

This chapter invites you to step behind the curtain. To see frequency not as a pre-existing property of reality—but as an emergent property of perception, built entirely on the scaffolding of observation, time, and interpretation.

Let's be clear:

- Vibrations exist.
- Energy structures shift.
- Fields resonate.

But frequency—as a quantitative label—only emerges when a conscious or designed system:

1. Observes a vibration,
2. Defines a time interval,
3. And counts how many recognizable cycles occur.

This is not semantics. This is structure. You don't discover frequency—you instantiate it through attention.

Why This Matters

This changes everything about how we understand light, time, and even the so-called photon.

Because if frequency is a derivative of observation, then:

- It cannot exist before observation.
- It cannot be transferred like a physical object.
- And it cannot be used to define an independent particle (like the photon) without circular logic.

Instead, frequency becomes the bridge between vibration and realization between the universe's activity and the observer's framework.

Chapter Premise:

Frequency is not a fixed property of reality. It is a relational outcome—defined when an observer enters the frame and organizes vibration within time.

In the sections ahead, we will explore this premise through UED, revealing how time emerges, how observation activates measurement, and why frequency is not what happens in the world—but what happens in you, when you engage with the world.

4.1 Time as an Emergent Frame

If frequency is "how often something happens in a given time," then time must come first. Right?

That's what classical physics has told us for centuries.

Time—according to Newton—was absolute. A universal clock ticking away in the background, independent of matter, space, or consciousness. All change was assumed to occur within time, like actors moving across a pre-constructed stage.

Einstein challenged this by tying time to motion, frames of reference, and gravity—but even in relativity, time remains a dimension, a coordinate on the map of spacetime, mathematically precise and physically real.

But Universal Energy Dynamics (UED) asks a more foundational question:

Is time something that exists on its own? Or is it something that emerges when structured energy begins to interact meaningfully?

UED View: Time Is Not a Background—It's a Consequence

In UED, time is not an empty canvas upon which reality paints its events. Instead, time is a relational ordering of change—a byproduct of interactions between energetic structures.

Said differently:

- Where there is no interaction, there is no need for sequence.
- Where nothing is observed, nothing is timed.
- Time becomes meaningful only when energy begins to restructure, and those restructurings are measured or realized by a system capable of tracking difference.

It's not the ticking of a cosmic watch. It's the recognition of variation by an observer—biological, mechanical, or conscious.

No Observation, No Time

To understand this more intuitively, imagine a universe:

- With no clocks, no observers, no systems to record or respond.
- Just structured energy, stable, vibrating silently across infinity.

Does time exist in that universe?

In classical terms, yes. "Time flows anyway." But in UED, there is no frame against which to count. There is only potential transformation—motion without a defined beginning or end.

Now, introduce an observer—any structure that can:

- Detect variation
- Record sequence
- Compare states

Suddenly, the order of events emerges. This is not because time "arrived," but because the capacity to interpret energy sequences was activated.

Time is not real until it is realized.

Observation Defines Sequence

Let's be clear: UED does not deny that things change. It does not deny that vibration, motion, or fluctuation exist.

What it says is this:

The concept of "before" and "after"—and the ability to quantify frequency—only exist when a system begins to relate to those changes.

Time is not an independent entity. It is a perceptual scaffolding, born from the structured awareness of transformation.

Without observation:

- Change still exists, but time does not.
- Energy still fluctuates, but frequency cannot be defined.
- The universe vibrates—but without sequence, without clocks, without rhythm.

Time is not the canvas. Time is the brushstroke of interpretation, painted onto vibrating energy by the structure of observation.

Conclusion

In UED, time is not a container—it is a condition. It does not tick. It emerges. It does not flow. It organizes.

And once we understand this, we realize:

Frequency is not the property of a wave. It is the outcome of a relationship between wave and observer—structured through time that the observer itself defines.

4.2 No Observer, No Clock

Let's begin with a provocative claim:

There is no such thing as time without an observer. There are only vibrations—waiting to be noticed.

The statement may seem radical at first, but it unveils a deep truth at the heart of frequency, perception, and the illusion of temporal continuity.

We often treat clocks as *independent validators* of time. We imagine them as machines that "tell" time, even if no one is watching.

But what does a clock *really* do?

Clocks Don't Measure Time—They Mirror It

Whether it's a quartz wristwatch, a pendulum, or an atomic clock, every timekeeping device is essentially a resonant system. It is built upon:

- A repeating event (like the oscillation of a crystal or atom),
- A reference frame (a mechanism or memory that counts

those oscillations),
- And a relationship to observation (human or instrumental).

Without that relationship—without someone or something *interpreting* the repetition—there is no meaningful measurement of time.

A clock on a deserted planet, ticking endlessly with no receiver, does not "create time." It only cycles structure—potential rhythm that becomes perceived time only if another system enters and syncs with it.

Observation Assigns Order

An observer—human, technological, or otherwise—is not passive. The act of observation is structure in action:

- It notices change.
- It orders sequences.
- It tracks "before" and "after."

Without observation:

- No event is marked.
- No interval is counted.
- No "second" is born.

The universe could vibrate eternally, but if no system experiences those vibrations in relational sequence, then there is no time—just unrealized structure.

All Clocks Are Resonance Tools

Let's look closer.

Every clock is built upon a natural resonance:

- A quartz crystal vibrating at ~32,768 Hz.
- A cesium atom flipping its state at 9,192,631,770 Hz.
- A heartbeat pulsing in rhythm with a metabolic loop.

But these systems don't *know* they are clocks. They only become "clocks" when another system interprets their rhythm.

Your watch doesn't "know" a second passed. You do—because your awareness has synchronized with the tool's oscillation. Your mind gives structure to flow.

Thus, clocks don't tell time.

Clocks resonate. You decide what it means.

No Observer → No Before or After

Imagine a vibration happening in deep space. A wave. A pulse. A fluctuation in the vacuum.

If there is no structure nearby to measure it, track it, or respond to it, then:

- There is no start or end.
- There is no sequence.
- There is no frequency—because no frame of time was ever invoked.

This is the essence of UED's view:

Observation is the trigger that activates time perception.

Only when a system enters the field and synchronizes with a pattern does temporal meaning emerge.

That meaning isn't an external truth—it is a realized structure, a subjective ordering we impose on vibration.

Frequency Emerges Inside a Temporal Loop

And here's the crucial consequence:

Since frequency is defined as cycles per second, and a second is a unit *created by an observing system*, then:

Frequency doesn't exist until the loop of observation is closed.

There must be:

- A system capable of recognizing repetition.
- A structure for measuring duration.
- And an awareness—conscious or computational—that can perceive difference across moments.

Only then do vibrations gain frequency. Only then does reality become rhythm. Only then does energy become measurable through time.

Final Insight

We don't find frequency in the wave. We assign frequency to the wave once we observe it within a defined time frame.

Without the observer, there is no clock. Without the clock, there is no frequency. And without frequency, the photon model collapses—not for lack of energy, but for lack of realization.

4.3 Why Frequency Is a Product, Not a Precondition

In mainstream physics, frequency is often treated as a fundamental attribute—something inherently "belonging" to a wave or vibration.

A wave of light? It "has" a frequency. An atom oscillating? It "produces" frequency. A signal emitted from a satellite? It "carries" frequency.

This view is so ingrained that we rarely pause to question it. But Universal Energy Dynamics (UED) invites us to look deeper.

And when we do, we discover something profound:

Vibration is fundamental. Frequency is not.

Vibration Exists Without Being Counted

Before anything can "have" frequency, it must first vibrate—that is, exhibit structured motion or recurring energy dynamics within a field.

These vibrations:

- Are continuous in nature.
- Occur without needing an observer.
- Are intrinsic to how energy seeks equilibrium within the field.

But frequency—the familiar "number of cycles per second"—only comes into existence when an observer defines a unit of time and counts repetitions within it.

This means:

- Vibration is real.
- Frequency is constructed.

The moment you say, "this wave has a frequency of 500 Hz," you are not describing what is, but what has been measured and interpreted.

UED Flips the Logic

In classical models:

- Frequency is a precondition—an innate, defining characteristic of the wave.
- It travels with the wave. It belongs to the vibration.

In UED, the order is reversed:

Vibrations are raw expressions of energetic motion. Frequency is a numerical outcome—a ratio we assign once we observe those vibrations within a chosen time frame.

It's not part of the vibration. It's a reflection of how we've chosen to interact with it.

Measurement Produces Frequency

Let's be clear: A wave doesn't *have* 440 Hz until someone defines a "second" and counts 440 cycles within that span.

Just as a mountain doesn't *have* a "height" until you define a starting point and a vertical metric,

a vibration doesn't *have* a "frequency" until you define:

- A time window (e.g., one second),
- A frame of reference (e.g., inertial or oscillating),

- A counting mechanism (e.g., sensor, clock, eye, brain).

This is why:

The act of measuring creates frequency—just as the act of measuring creates distance in a curved field.

We don't discover frequency like an artifact. We assign it as a contextual metric within a system of interpretation.

Frequency Is an Interpretive Abstraction

Let's emphasize the difference again:

Vibration	Frequency
Continuous, raw energy motion	Quantified cycles per chosen time unit
Exists without observation	Emerges only through observation
Intrinsic to structured field behavior	Extrinsic, applied by a measuring system
Can happen with no start or stop	Requires temporal framing to be defined

This reframe changes our understanding of wave behavior entirely.

A photon isn't born with a frequency. What we call "photon frequency" is what *we have determined* by observing how our system resonates with the underlying vibration.

Final Thought

Frequency is not found—it is formed. Not given by nature, but granted by structure.

This doesn't make frequency less real—but it does make it relational, not absolute.

And once we accept this, the implications ripple outward:

- The photon cannot carry energy "because of frequency."
- The photoelectric effect doesn't "prove" frequency exists before observation.
- Energy is not "quantized" in space—it is realized in structured observation.

Which leads to the next insight:

If frequency is emergent, then so is the *photon*. Not a particle—but a pattern made meaningful by observation.

In the next section, we'll explore this realization fully.

4.4 Vibrations Exist—Frequency Is Our Interpretation

Now that we've dismantled the assumption that frequency is a pre-existing property of waves, let's bring the argument full circle:

What exists universally is not frequency—but vibration.

Vibration is fundamental. It is the expression of energy in action—a manifestation of disturbance seeking equilibrium across the structured field of existence.

It is what the universe *does*, endlessly and everywhere. From the dance of atoms to the warble of galaxies, vibration is the language of energy—oscillating, radiating, modulating itself through space and potential.

But Frequency? That's Our Lens.

What we call "frequency" is not written into the wave like an internal code. It's not branded onto the photon or etched into the string of some vibrating particle.

Frequency is the pattern that emerges when a system—capable of tracking, comparing, and structuring time—engages with vibration.

And that system... is the observer.

Whether it's a human brain, a digital sensor, or an atomic oscillator—every act of frequency detection is really an act of interpretation.

We take:

- A continuous motion,
- Apply a temporal filter,
- And count: "This many cycles per that many seconds."

That count is not the wave. It is the way we experience the wave.

The Observer's Role: Central, Not Passive

Let's restate it boldly:

Classical Assumption	UED Realization
Frequency is a property of the wave	Frequency is a construct of observation
The observer detects what is already there	The observer co-defines what becomes visible
Energy is transferred through cycles	Energy is realized when resonance is framed

This changes everything.

Because it means:

- Light doesn't "have" frequency.
- Particles don't "carry" frequency.
- Photons don't "arrive" pre-coded with energy.

Instead:

Structured vibration travels. Frequency emerges when an observer enters the system and gives time to motion.

Why This Matters

We often assume our measurements are windows to truth. But in reality, they are mirrors of interpretation.

The frequency you read off an oscilloscope is not something *in the wire.* It's the way your system—made of circuits, clocks, and code—has chosen to interpret recurring motion.

The color you see when you look at light isn't a wavelength printed onto the wave. It's the way your retina + brain system processes structured vibrations into a perceptual scale.

And the photon frequency you use in quantum physics? It's not an internal attribute—it's a relational value, describing how that vibration resonates with your measurement system over time.

Final Insight: You Don't See Frequency—You Assign It

So, let us conclude with clarity:

Vibration is real. Frequency is relative.

The wave doesn't hum at 500 trillion times per second. You count 500 trillion patterns because your system chooses to slice time that way.

In the absence of an observer:

- No slice.
- No rhythm.
- No frequency.

Just pure vibration, flowing freely through structured vacuum, always seeking balance—always waiting to be witnessed.

And when you do witness it?

You call it a wave. You measure a frequency. You see a photon. But what you really saw... was your own resonance meeting the universe's vibration.

4.5 The Resonant Signature

Let's pause and collect what we've just dismantled—and what we've just discovered.

We began this chapter with a simple statement:

"You don't observe frequency. You define it."

From there, we uncovered that:

- Time is not absolute—it arises through structured interaction.
- Frequency is not fundamental—it emerges when observation imposes rhythm.
- Measurement doesn't reveal vibration—it interprets it.

So if all of this is true, then the most revered concept in quantum light theory must be revisited.

The Key Insight: What Are Photons Really?

In conventional physics, the photon is defined as a quantized packet of light, carrying energy proportional to its frequency via:

$E=h\nu$

But we've now seen that:

- Frequency is not intrinsic—it is assigned through observation.
- Energy in this context is not transferred—it is realized through resonance.
- Quantization is not built into space—it's a product of structural limits of alignment.

So then the unavoidable question becomes:

If frequency arises from observation, and photons are "quantized" based on frequency, then what exactly is a photon?

Is it:

- A flying massless particle?
- A self-contained burst of energy?
- A pre-packaged cosmic messenger with its own identity?

Or is it something much more elegant, much more accurate—and far less physical?

From Particle to Pattern

In the light of UED, the photon is not:

- A particle traveling through space.
- A quantized bullet of energy.
- A courier delivering packets from point A to B.

Instead, it is:

A resonant signature. A realized pattern born from the interaction between a universal vibration and an observing structure.

It is not a thing. It is an intersection.

A photon "appears" not because it *arrives*—but because a resonance condition is met:

- The source is vibrating.
- The field is structured.
- The observer's system is tuned.

And when those three align?

We say, "Ah, a photon!" But what we've really observed is a moment of harmony—a match between field disturbance and system resonance.

Preparing for next chapter: The Resonant Signature

In the next chapter, we will let go of the photon as an entity and fully embrace it as a phenomenon—the realization of structured energy at the point of resonance.

We will explore:

- How vibrational signatures behave like particles without being particles.
- Why quantization is not a constraint of nature—but a result of resonance thresholds.
- How light becomes "visible," "measurable," and "useful" only when structure, intent, and observation form a triadic lock.

We're about to redefine what light actually *is*—not just in scientific terms, but in structural reality.

Not a wave. Not a particle. Not a duality.

But a resonant event—a flash of coherence between the field and the observer.

Are you ready?

Let's meet the photon you've always known, but never truly seen—not as an object, but as a signature.

Key Takeaways from Chapter 4: Frequency Emerges with Observation

Frequency is not a built-in property of waves—it's a product of observation.

Vibrations exist universally, but what we call "frequency" only arises when an observer applies a time frame and counts repetitions. Without observation, there is no measured frequency—only raw, structured vibration.

Time itself is not absolute—it is emergent.

In UED, time is not a pre-existing flow but a relational sequence that emerges when structured systems (observers) begin organizing energetic changes into before-and-after patterns.

No observer means no clock—and no cycles to count.

Clocks, whether atomic or mechanical, do not generate time; they resonate. Time is created when a system aligns with that resonance and uses it to organize experience.

Frequency is an interpretive abstraction, not an intrinsic identity.

Vibrations are continuous. Frequency is the numeric overlay we assign through temporal framing. It is not a feature of the wave—it is a result of the observer's perspective.

The act of measuring constructs the phenomenon.

Just as distance is meaningless without a ruler and curvature distorts measurement, frequency only exists because we frame vibrations within artificial temporal loops.

The observer is central to all realization.

Whether it's time, frequency, or the appearance of light itself, nothing "exists" in a quantifiable sense until observation completes the resonance loop between source, medium, and receiver.

What we perceive as light frequency is simply structured vibration aligned with our system.

A wave doesn't hum at 600 THz unless your perception system slices time and counts cycles that way. Frequency is not "carried" by the wave—it is *how you choose to hear it.*

This redefinition demands a new model of light.

If photons are defined by frequency, and frequency arises through observation, then the photon is not a particle—it is a **resonant signature**, a moment of vibrational alignment we interpret as quantized light.

Chapter 5: $E = h\nu$ – The Observer's Equation

"A Beautiful Equation, But Not a Universal Truth"

In the world of modern physics, few equations are quoted with such reverence—and such unquestioned authority—as:

$E=h\nu$

It is elegant. It is concise. And it changed everything.

This single formula, proposed by Max Planck and later popularized by Einstein, is said to define the energy of light. A beam, a pulse, a particle—a photon—is said to carry an amount of energy proportional to its frequency. That energy, we are told, is the product of a universal constant h (Planck's constant) and a measurable frequency ν.

Textbooks present it as a law. Scientists rely on it as a foundation. And every photon, it seems, must obey it.

But what if this equation is not universal?

What if it's not telling us something about the photon, but something about ourselves?

What if $E=h\nu$ is not a truth of nature—but a truth of resonance, emerging only when the observer enters the system?

Cultural and Scientific Significance

Before we question the sacred, let us honor it.

Planck's equation was born in 1900—not as a philosophical declaration, but as a mathematical necessity. Planck needed to explain blackbody radiation, and the classical equations of the time were failing.

By assuming that energy could only be emitted or absorbed in discrete packets—quanta—Planck fixed the equations. The concept of frequency-dependent energy worked.

Einstein would later seize this idea to explain the photoelectric effect, claiming that light itself was quantized: each photon carried energy proportional to its frequency.

Together, these two moments formed the backbone of quantum physics.

They gave rise to:

- The photon as an entity,
- The quantization of the electromagnetic field,
- And the probabilistic mechanics that dominate our understanding of the subatomic world.

To this day, $E=h\nu$ appears on the walls of classrooms, in quantum optics, in laser design, in semiconductor theory, in astrophysics—it is, quite literally, everywhere.

But ubiquity should not be confused with universality.

The Crux: What Are We Really Measuring?

As we've seen in the previous chapter:

- Frequency is not fundamental—it emerges through observation.

- It requires time, which itself is a framework induced by systems capable of sequencing change.
- Therefore, the value of ν in Planck's equation is not an objective feature of a wave—it's an observer's numerical assignment.

So then, what exactly is E, the energy we're calculating?

Is it:

- Energy carried by a particle through space?
- Energy objectively stored in a field?
- Or energy realized at the moment of resonance, when vibration meets structured observation?

Reframing the Inquiry

This is the question at the heart of this chapter:

If frequency is observer-defined, then the energy we calculate from it is also observer-bound. So is $E=h\nu$ a law of the universe—or a law of resonance?

We do not deny the accuracy of the equation. We honor its usefulness, its predictive power, and its role in technological advancement.

But in UED, we reposition it.

Not as a cosmic constant—but as a local realization. Not a declaration of what is—but a reflection of what has been measured through alignment.

In the next section, we'll revisit the origin of this equation, not as myth, but as a moment of mathematical pragmatism—a brilliant patch for a broken model, now ready to be reinterpreted through a deeper understanding of structure, vibration, and observer-induced reality.

5.1 Planck's Equation in Context

To understand what E=hν truly represents, we must return to the moment of its birth—not as a universal proclamation of nature, but as a mathematical necessity born out of crisis.

The Blackbody Radiation Puzzle

At the end of the 19th century, physicists faced an embarrassing problem. According to classical electrodynamics, when you modeled the radiation emitted by a perfect thermal emitter—a blackbody—the energy should increase infinitely as the frequency increased. This prediction became known as the ultraviolet catastrophe.

Reality, however, refused to cooperate. Empirical data showed that:

- Radiation peaked at certain frequencies.
- It diminished at higher frequencies—exactly the opposite of what theory predicted.

No amount of classical tweaking could fix it.

Enter Max Planck.

Planck's Radical Assumption: Energy Comes in Packets

To make the equations match the experimental data, Planck did something extraordinary—not because he believed it philosophically, but because it worked mathematically.

He proposed that:

Energy isn't emitted continuously, but in discrete packets—quanta—proportional to the frequency of oscillation.

That is:

$E=h\nu$

Where:

- E is the energy of one quantum of radiation,
- h is Planck's constant,
- ν is the frequency of the oscillator (not the wave itself yet).

At the time, Planck treated this as a fitting tool, not a description of nature. He was trying to save the mathematics, not rewrite the universe.

In fact, Planck himself remained skeptical of the physical interpretation of quantization.

He introduced the idea with caution, as a mechanism that "just happened" to work.

Einstein's Leap: From Oscillators to Photons

Then came Einstein, who made the bold leap.

In 1905, while studying the photoelectric effect, Einstein used Planck's relation in a new context. He assumed that light itself—not just the material emitters—was quantized.

He postulated that:

- Light consisted of particles, later called photons, each carrying energy $E=h\nu$.
- A single photon could collide with a single electron and eject it—but only if its frequency (and thus energy) exceeded a threshold.

Einstein's genius lay in applying Planck's relation to an observer-interaction experiment, not a thermal model.

And it worked.

Electrons were ejected instantly at the threshold frequency, regardless of intensity. Planck's mathematical fix became a physical mechanism—one that made Einstein famous and helped ignite the quantum revolution.

But What Were They Really Describing?

Here's the critical point:

Neither Planck nor Einstein claimed that $E=h\nu$ described an object traveling through space.

Planck saw it as a tool to describe how energy is emitted. Einstein used it to model how energy is absorbed.

In both cases, the equation was applied in observer-bound setups:

- Detectors.
- Receivers.
- Measurement systems.

They were dealing not with reality as it exists independently, but reality as it becomes observable—in controlled, structured contexts.

So, what is E=hν actually describing?

Not a particle. Not a wave. Not an entity flying through the vacuum.

It is describing the amount of energy that becomes *realized* when a vibration matches the resonant structure of an observing system.

Reframing the Equation Through UED

In Universal Energy Dynamics, E=hν is not wrong. It's just misinterpreted.

We don't discard it—we reassign its role.

- It doesn't describe the "contents" of a photon.
- It doesn't describe what light *is* in transit.

Instead:

It describes the energy realized at the point of resonance, when structured vibration becomes measurable through time-defined observation.

This realization is not an inherent property of the wave. It's a relational phenomenon—a consequence of alignment between:

- Field structure (vibration),

- Observer framework (time/frequency),
- And resonance bandwidth (material threshold).

Only when those align, does $E=h\nu$ become "true." Outside that window, the equation has no active meaning—it becomes a dormant potential, not a physical fact.

The Shift We Now Face

So as we move forward, we must ask:

If energy is only revealed through resonance, and frequency is only defined through observation, then what exactly is being "quantized"?

That question leads us directly to the heart of the chapter—and to the next section

5.2 Why It's a Local Truth, Not a Universal Law

Throughout the history of physics, equations have acquired an almost sacred status. The cleaner, more elegant an equation looks, the more we tend to assume it must represent some deep, objective truth of the universe.

Few equations have enjoyed this kind of reverence more than:

$E=h\nu$

To generations of physicists, this relation is not just useful—it's fundamental. It's said to define how nature parcels out energy. To describe how light exists. To quantify what photons are.

But this trust—this assumption of universality—deserves scrutiny.

Because in reality, E=hν is not a law of how the universe works everywhere, for everyone, under all conditions.

It is a local rule that becomes true only when an observing system defines frequency, and resonance occurs.

And once we see this, the equation reveals something even more beautiful than universal constancy—it reveals contextual realization.

From Law to Locality: The Observer's Role

In classical thinking, equations like E=hν are thought to describe inherent properties:

- Light has a frequency.
- It carries energy.
- That energy exists even if no one is there to measure it.

But Universal Energy Dynamics (UED) reframes this drastically:

Frequency is not intrinsic—it is defined by the observer. Therefore, energy—when tied to that frequency—is not carried. It is realized.

Let's make this clear:

- The equation doesn't describe something flying through space.
- It describes a measured outcome—the energy *extracted* when a structure resonates with vibration over time.

Without an observer:

- There's no frequency to apply.
- And no energy to measure.
- Only structured vibration waiting for a system to bring it

into realized form.

The LED Example: Quantization by Alignment

Consider a modern example: a white LED.

- It emits a broad spectrum of structured vibrations—ranging across many wavelengths and intensities.
- These emissions radiate outward, whether someone is looking or not.

Now introduce a tuned receiver—say, a photodiode designed to detect light at 520 nm (green).

What happens?

- The receiver doesn't *see* the entire emission.
- It responds only to the part of the vibration that matches its own internal resonance band.
- It "registers" that portion of the vibration as a discrete energy event.
- And it does so by assigning a frequency and applying $E=h\nu$

But here's the insight:

The LED didn't emit a "photon" with that frequency. The observer selected and quantized that structured vibration based on its own architecture.

This is what we mean when we say:

Quantization is not what light carries—it's what the system cuts out and realizes.

Local, Not Global

This leads to a profound realization:

- $E = h\nu$ is not universally descriptive.
- It's locally activated—a result of resonance, not existence.

It works:

- When a vibration is observed,
- When time is applied,
- And when a system's sensitivity frames the wave into a countable form.

Outside of those conditions, energy is not lost, but it remains unrealized—not because it's not there, but because the conditions for alignment were not met.

The field is full of potential. The light is full of vibration. But until observation structures it, there is no measurable "photon," no calculated energy, no meaningful frequency.

Summary Insight

So we return to the central premise:

Planck's equation is not a universal law written into the fabric of the cosmos. It is a local truth—revealed when resonance, structure, and observation complete the circuit of realization.

It does not describe what light *is*. It describes what light *becomes* when we interact with it.

Up next: we deepen this distinction by exploring why resonance does not guarantee ontological reality—and how treating resonance as "truth" can lead to foundational misinterpretations in physics.

5.3 Resonance ≠ Reality

Resonance is often mistaken for revelation. The moment we measure something cleanly, repeatedly, and predictably, we tend to assume: "That's how it really is."

But clarity is not confirmation. Repeatability is not reality. And resonance—no matter how precise—is not proof of ontological existence.

Resonance is a condition of alignment. It tells you that a system responded—not that a thing inherently existed.

The Musical Analogy: Notes Don't Exist Until Played

Imagine a violin string. Left untouched, it holds tension, structure, and potential. But no sound.

Now imagine drawing a bow across it.

What do you hear?

A note. A tone. A beautiful, quantifiable frequency. 440 Hz? That's an "A" in Western music.

But here's the twist:

- The 440 Hz isn't *in* the string.
- It's not *inside* the violin body.
- It's not a particle waiting to escape.

That frequency emerges when:

1. The string is activated,

2. The body of the violin amplifies the vibration,
3. Your ear (or an instrument) resonates with the pattern and assigns time to it.

The note didn't exist before this triad of interaction, amplification, and observation. What existed was structure—waiting to be harmonized into recognition.

The Universe Is Like That Violin

Light, energy, vibration—they're all around. But the universe doesn't "send" photons. It doesn't carry around packets labeled with frequencies.

What the universe has is structured energy—vibrations in fields, potential in equilibrium, patterns waiting to be engaged.

And what we experience as:

- Frequency,
- Energy,
- Quantized photons,

...are the musical notes that emerge when our instruments resonate with that structure.

Resonance Is Relational—Not Intrinsic

In philosophy, there's a critical distinction between:

- Ontology – what *is*, independent of perception
- Epistemology – what we *know*, based on observation

Resonance belongs to the second category.

Just because your system resonates at 500 THz doesn't mean that 500 THz "exists" in the vacuum.

Just because you observe a photon with energy $E=h\nu$, doesn't mean a particle with that energy was traveling through space.

What you've observed is alignment between:

- A vibrational field,
- A structured detector,
- And your assigned time window.

In that context, yes, energy appears. But it appears for you, not because it was traveling independently waiting to be found.

Resonance as Collapse

The act of measurement is not a passive discovery—it's a collapse of possibility into visibility.

- In quantum physics, we speak of wavefunction collapse.
- In music, we hear silence until the note is struck.
- In UED, energy is diffuse until resonance structures it into form.

That's what $E=h\nu$ captures—not a pre-loaded photon, but a moment of realization where universal vibration becomes a localized energetic signature.

Not because the energy was flying through space. But because a structure was struck—in harmony with your own.

Final Thought

Resonance doesn't prove reality. It constructs a shared moment between structure and perception.

The universe does not contain frequency. It expresses vibration.

The universe does not carry energy. It structures potential.

And only when your system tunes itself to this structure—only then does something emerge that you can call a photon, a note, a measurement, or a law.

Up next: We translate this realization into a governing principle of observation—Chavan's Law of Frequency—which will crystallize the core philosophy of this entire framework.

5.4 Chavan's Law of Frequency

After exploring the emergent nature of time, the constructed identity of frequency, and the observer-bound realization of energy, it's time to crystallize these insights into a governing principle—a law not of matter, but of meaning.

This is where we introduce:

Chavan's Law of Frequency:

"Frequency is not what vibration carries. Frequency is what the observer extracts."

This law reframes one of the most fundamental assumptions in physics—that frequency is an inherent trait of a wave, encoded into the universe like a digital watermark. Instead, it asserts:

- Vibrations are real, yes.
- But frequency is not a delivered truth—it is an observed construction.
- The moment you assign a number of cycles per second, you are already engaged in interpretation, not description.

The Observer as Co-Creator

This law positions the observer—not as a passive detector—but as an active participant in the formation of physical knowledge.

- You don't just discover frequency; you construct it through your own structural filters (biological, mechanical, computational).
- You don't just receive energy; you realize it by collapsing a wave's potential through resonant interaction.
- Your instruments—eyes, detectors, sensors—are tuning forks, not telescopes. They ring with reality, but never show it in its untouched form.

In this sense, every measurement is not a reading—it is a relational event, a structured response between you and the universe.

E = hν: A Consequence, Not a Cause

With Chavan's Law in place, we can now reinterpret the most celebrated energy-frequency relation of all:

$E=h\nu$

No longer as a universal law of nature, but as a conditional expression of resonance.

- It is valid only when an observer (or observing system)

defines a time frame.
- It is realized only when vibration aligns with a resonant structure.
- It describes not what light inherently *is*, but what becomes measurable *when structure meets structure.*

Thus, Planck's equation becomes a bridge, not a boundary. A translation rule, not a fundamental truth.

Planck's Constant: h as a Scaling Bridge

In this new framework, even the revered Planck constant hhh gains a more refined role.

- It is not the "quantum of action" as a fundamental building block.
- It is the scaling ratio between two domains:
  - Unstructured vibration in the field, and
  - Quantized realization at the point of observation.

It links:

- The continuous to the discrete,
- The field to the frame,
- The vibration to the number.

h doesn't quantify energy—it mediates resonance.

And the moment you change the observing structure—biological vs. mechanical, analog vs. digital, slow vs. fast—the experience of energy, and the role of h, transforms accordingly.

Implications of Chavan's Law

- Physics becomes contextual: not a map of fixed truths, but a library of resonant outcomes across structured observations.
- Light becomes participatory: not a signal delivered, but a signature extracted.
- Science becomes creative: not merely descriptive, but generative—co-authored by the universe and the observer's structure.

And perhaps most importantly:

The photon—the so-called "quantum of light"—is no longer a particle in transit. It is a pulse of alignment, a resonant event between field vibration and observational bandwidth.

Leading into the Next Chapter:

Now that we've redefined frequency as emergent and energy as relational, we're ready to dismantle the illusion of transmission itself.

What is light between observations? Where does it "go" when no one is there to detect it? If nothing is "emitted," and nothing is "traveling," then what exactly is illumination?

5.5 Transition to Next Chapter – Seeing Light Without Carriers

Now that we've reframed $E=h\nu$ as a localized expression of resonance—not a universal law but an observer-induced realization—a profound new question arises:

If energy is not carried, but realized... If frequency is not fixed, but emergent...Then what is light *between* resonance events?

Is it:

- A wave moving through space?
- A particle in transit, like a courier en route to a destination?
- A photon traveling through vacuum, waiting to be absorbed?

Or have we misunderstood light's journey entirely?

The Illusion of Transmission

We've been taught to imagine light as something that moves— a packet of energy emitted from a source, traversing a distance, and finally arriving at a receiver.

But now, armed with Chavan's Law and the UED framework, we are compelled to reconsider:

- The photon is not a packet—it is a resonant signature.
- Frequency is not inherent—it is a relational output of observation.
- Energy is not sent—it is realized through alignment.

So what is light doing in the space between source and receiver?

The radical answer is: Nothing at all.

Between Resonances, There Is Only Potential

In the absence of a resonant structure (an observing system), light does not exist as a traveling object.

- No wave propagates with "frequency."
- No photon moves with "energy."
- No signal carries "information."

What exists instead is structured vibration in the field—present, but unrealized.

This is not emptiness. This is potentiality—energy waiting to be shaped into form through resonance.

Light is not traveling through space. It is waiting to be realized within structure.

Shifting from Motion to Realization

This shift—from movement to moment—transforms how we understand everything from vision to information transfer, from astronomy to quantum theory.

- The source is vibrating.
- The field is structured.
- The observer defines time and filters resonance.
- And when alignment happens—light appears.

Not because it has *arrived*, but because you have tuned into it.

Light as Realization – Nothing Travels, Everything Aligns

In the next chapter, we will fully dismantle the myth of light-in-motion and introduce a more precise framework:

- Light as a condition of resonance, not a product of emission.
- Observation as the catalyst, not the destination.
- The field as ever-ready, with no need for delivery—only alignment.

We will explore how:

- Communication happens without transfer.
- Illumination is an emergent experience.
- And photons... may never have moved at all.

Because when vibration, field, and observation resonate, light is not received—it is realized.

Key Takeaways from Chapter 5: E = hν – The Observer's Equation

Planck's equation E=hν was never meant as a universal law.

It was originally a mathematical fix for blackbody radiation and later used by Einstein to model energy exchange in specific, observer-bound experiments—not to define objective particles in motion.

Frequency is not inherent to vibration—it is observer-defined.

A wave does not carry frequency. Frequency arises only when an observer applies a temporal frame and extracts cycles per second. Without observation, there is vibration—but no frequency.

Energy is not transferred—it is realized.

The energy described by E=hν is not something "carried" by a photon. It is brought into existence at the moment a structured vibration aligns with a system capable of recognizing and measuring it.

Resonance is relational, not absolute.

The appearance of quantized energy is a result of resonance between structures—not a reflection of fixed, traveling entities. Measurement is not discovery; it is alignment.

Planck's constant h is a bridge—not a boundary.

Rather than being a universal building block, hhh functions as a scaling constant, converting unstructured field vibration into discrete energy events—only when observer-defined conditions are met.

Chavan's Law of Frequency reframes the photon paradigm:

"Frequency is not what vibration carries. Frequency is what the observer extracts."

This principle reveals that photons, energy quanta, and the very act of measurement are all emergent—not fundamental.

The photon is not a particle—it is a resonant signature.

What we call a photon is the product of field vibration intersecting with observation. It doesn't travel; it materializes when structure meets structure, forming a measurable event.

$E = h\nu$ is not a cosmic rule—it's a localized realization.

This equation describes what energy becomes for us, not what energy inherently is. It is a powerful tool of interpretation, not a window into an independent reality.

Chapter 6: Resonance

The Real Cause Behind All Effects

"What doesn't resonate, doesn't exist for you."

Take a deep breath. Look around. Everything you notice, everything you feel, everything you believe is *there*—is only there because you resonated with it.

A wave of light travels past you? If your eyes don't align with its frequency, you never see it. A song plays in a distant room? If it doesn't reach your ears' sensitivity band, it's as good as silence. An idea is spoken in a lecture? If your mind doesn't match its structure, it passes by like static—unfelt, unrealized.

"If it doesn't resonate, it doesn't register. If it doesn't register, it doesn't exist—for you."

This is not metaphor. This is structure in action.

Resonance Is More Than Physics

In classical physics, resonance is often reduced to a mechanical phenomenon:

- A pendulum swinging in sync with another.
- A tuning fork vibrating sympathetically.
- A bridge collapsing under oscillatory force.

But that's just the outer shell of a much deeper principle.

Resonance is the real cause behind what we perceive as effect.

Not impact. Not transmission. Not delivery.

Resonance is the moment when the universe and observer vibrate in structural agreement—when something dormant becomes realized.

Whether it's:

- Light meeting retina,
- Sound meeting eardrum,
- Insight meeting thought,
- Emotion meeting memory—

Nothing truly begins until resonance completes a circuit.

You Don't Discover Reality—You Align With It

We've been taught to believe that the world *sends us signals,* and we just need to detect them.

But detection is not passive—it is a function of structure. Your brain, your instruments, your emotions—each is a filter that resonates with certain inputs and ignores the rest.

This is why:

- The same light can be seen by one creature and not another.
- The same sentence can enlighten one person and confuse another.
- The same moment can inspire one soul and pass unnoticed by another.

Reality, then, is not what's out there. Reality is what *resonates* in here.

From Light to Thought—Resonance Rules Everything

Physics tells us energy moves. But UED and Chavan's Laws reveal something deeper: energy aligns.

The act of seeing, knowing, feeling, or believing—is not caused by what's sent. It's caused by what resonates.

This chapter will now explore how resonance:

- Governs perception,
- Shapes illumination,
- Structures consequences,
- And redefines energy as realization, not transfer.

Because at every level—from atoms to awareness—resonance is the true cause behind all observed effects.

6.1 What Resonates, Responds

In traditional physics, we explain events through force and motion.

A cue ball strikes another—it moves.

A charge applies potential—current flows.

A photon hits an electron—it gets ejected.

Cause and effect seem simple: apply force → get response.

But that model begins to fall apart at the edges—especially when you go deeper into quantum behavior, consciousness, or even communication.

Why?

Because not everything responds to force. But everything responds to resonance.

Examples That Prove the Point

1. A Radio Receiver Turn a radio on and sweep through the dial. You won't hear all frequencies at once—you'll only receive the station you're tuned to. Not because others aren't "present," but because your system isn't structured to match them.
2. A Wine Glass and a Violin Note Play a violin near a delicate wine glass. Most notes won't affect it. But play its resonant frequency, and the glass may shatter. Force doesn't break the glass—structural alignment does.
3. The Photoelectric Effect Shine red light on a metal plate—nothing happens. Switch to ultraviolet—and electrons fly out. Intensity doesn't matter; frequency threshold does. Why? Because only certain structured vibrations align with the resonance band of that material.

Response Is Not Caused—It Is Triggered by Fit

These aren't exceptions. They are evidence of a universal pattern.

Systems don't respond because something *acts on* them. They respond when something matches them.

This isn't about energy overpowering a threshold—It's about vibration synchronizing with structure.

This brings us to a foundational truth:

Chavan's Law of Resonance

"Response is not the result of force—it is the realization of structural compatibility. What resonates, responds. What doesn't, remains unrealized."

This law flips classical cause-effect logic on its head.

Instead of saying:

- "X causes Y because it acted upon it,"

We now say:

- "Y was realized because X resonated with it."

From neurons firing, to emotional reactions, to the ejection of electrons, to thought awakening insight—it's all the same underlying process: alignment of internal structure with external vibration.

In UED: Resonance Is Consequence in Action

In the Universal Energy Dynamics framework, all fields are structured, and all systems are vibrational.

But interaction only happens when:

- The source vibration,
- The field structure, and
- The observer's frame

...find a match. When they do, a response is triggered—not because energy was *sent*, but because the conditions for realization were met.

Nothing is received. Nothing is pushed. Everything is *activated*—when resonance forms the bridge.

Consequence-Based Causality

This view aligns perfectly with the Law of Consequences (LOC):

Consequence is not merely reaction to action—it is the expression of energetic alignment seeking equilibrium.

- The radio doesn't play sound until your receiver aligns.
- The material doesn't release electrons until your light resonates.
- The mind doesn't awaken until the idea fits its inner rhythm.

Force alone cannot do this. Only structure meeting structure can.

Final Insight of the Section:

You don't change systems by pushing harder. You change systems by resonating deeper.

What resonates, responds. What doesn't resonate? It's not wrong. It's just not part of your realized reality—yet.

6.2 Perception as Alignment

What if seeing isn't receiving?

What if hearing, understanding, feeling, knowing—aren't about absorbing something "given," but about tuning into what's already there?

Perception has long been treated as a passive operation:

- The eye "captures" light.
- The ear "receives" sound.
- The brain "stores" information.

But this model misunderstands the true nature of perception.

Perception is not detection. Perception is resonance.

You don't see light because it was thrown at you. You see because your visual system aligned with the vibration. You don't hear music because sound entered your ears. You hear because your auditory structure resonated with the waveform.

Perception is not caused by signals. It is activated by compatibility.

Human Senses: Instruments of Resonance

Let's look at your biology—not as a receiver, but as a resonator.

- The eye is not a camera. It's a structure that responds only to certain wavelengths (approx. 380–740 nm). Anything beyond that—infrared, ultraviolet—passes through unacknowledged.
- The ear is a tuner. It vibrates sympathetically with sound waves between 20 Hz and 20,000 Hz. Below or above this, silence reigns—not because sound ceased, but because your structure no longer aligns.
- The skin responds to temperature, texture, pressure—but only through patterns of frequency in nerve impulses.

Your entire perceptual system is built not to "record" the universe, but to synchronize with it.

You do not sense everything. You sense what you can resonate with.

Technology: Extended Resonance

We built machines not to go beyond resonance—but to extend our structures of alignment.

- Telescopes don't "see farther." They translate faint vibrations into a resonant frame our eyes can interpret.
- Radio receivers don't catch invisible waves—they resonate selectively and produce measurable responses.
- Infrared cameras, UV detectors, X-ray scanners—each one is a specialized resonance filter that makes unseen vibrations available through structured alignment.

We have never built a device that detects everything. We have only ever built tools to expand the range of our resonance.

The Mind: A Resonant Processor of Meaning

Even your thoughts obey the same principle.

- You don't "get" ideas unless your cognitive structure aligns with the pattern.
- The same sentence spoken to a hundred people produces a hundred different realizations—why? Because meaning is not transmitted. It is resonated into.

Insight doesn't happen because something was taught. It happens when something in you aligns with the structure of what was presented.

Education is not information delivery. It is resonance facilitation.

Vision Is Not Reception—It Is Agreement

Now let's return to light.

We are told that:

- Light travels,
- Hits the retina,
- Triggers a signal.

But this is mechanical oversimplification.

In UED:

- The light field vibrates,
- The eye is structured to resonate,
- And vision occurs not because something was sent, but because alignment occurred.

You didn't receive a photon. You realized a vibration.

You didn't see light because it arrived. You saw because your structure said, *"I am ready to respond to this."*

Final Insight of the Section:

Perception is the phenomenon of structural alignment. Your biology, your technology, and your mind—all operate not as passive receivers, but as resonant gateways.

What you can't resonate with, you can't perceive. And what you can't perceive, doesn't exist in your reality. The universe doesn't choose what you see. Your structure does.

6.3 Illumination as Interaction, Not Delivery

Let's now challenge one of the most persistent myths in the human understanding of light:

The belief that light is something that *travels*, gets *delivered*, and then *arrives* at your eyes.

This model—of flying photons, streams of particles, and beams darting across space—has dominated both science education and cultural imagination.

But it's misleading.

Because what we call *illumination* doesn't happen when something arrives. It happens when something aligns.

Light as a Delivery System? Think Again.

Conventional thinking tells us this:

- A light bulb emits photons.
- Those photons move through space.
- They strike your retina.
- You "see."

This process sounds elegant—but it crumbles under deeper inspection:

- Where is the photon *before* you observe it?
- Does it *really* travel? Or is it part of a vibrating field you've only just noticed?

- And why does the same field produce *different experiences* for different observers?

Because the truth is subtler:

Illumination is not the arrival of energy. It is the activation of resonance.

From Sending to Synchronizing

Imagine two tuning forks. Strike one—and if the other is tuned to the same frequency, it begins to vibrate without being touched.

No energy was "delivered" across the air. There was no projectile. What happened was resonance.

Now scale that up.

- The source of light is vibrating in structured fields.
- The observer (biological or technological) has a particular structural bandwidth.
- When the field's pattern matches the observer's capacity—light is realized.

Not because something *came* to you, But because you were tuned to receive what was already there.

No Flying Photons, No Energetic Couriers

We imagine photons like messenger particles—tiny, energetic bullets zipping through space, bringing energy from one place to another.

But consider this:

- In quantum mechanics, the photon doesn't have a defined

position between emission and detection.
- In relativity, the photon experiences no time between those two events—it exists as a *boundary condition*.
- In UED, photons are not objects. They are resonant consequences.

There is no delivery van bringing light to your doorstep. There is only a field—and when you resonate with it, you call light into being.

Illumination Is a Closed Resonance Loop

Every moment of seeing is a completed circuit:

- A vibration seeking balance,
- A structure capable of receiving it,
- And a field that allows both to interact harmoniously.

That interaction is illumination.

It doesn't move. It occurs.

Like a key that unlocks a door, or a match that lights a flame—illumination is not transmission. It's transition.

From potential to presence. From field to form. From structure to sight.

Echo the Core Realization

Let's say it again:

Light is not seen because it arrives. Light is seen because resonance completes a loop.

That's why two people can stand side by side:

- One notices the light.
- The other doesn't.
- The light was "present," yes—but illumination occurred only where resonance was achieved.

This principle holds true:

- In physics (light),
- In sound (hearing),
- In emotion (connection),
- In cognition (insight),
- In consciousness (awakening).

The universe isn't sending you reality. It's waiting for you to tune into it.

6.4 The Law of Consequences in Frequency Behavior

Everything you've read so far—about perception, light, realization, resonance—can now be understood as not just conceptual alignment, but the energetic logic of the universe in action.

At the center of that logic is The Law of Consequences (LOC).

This is where physics meets purpose.

This is where structured energy, observation, and frequency stop behaving like abstract ideas and begin functioning as governed phenomena.

Every frequency you perceive, every burst of light, every realized signal—is not something the universe imposed on you. It is the consequence of your alignment with the field.

Frequency Is the Consequence of Structure Meeting Structure

Let's rephrase what we thought we knew.

You don't *observe* frequency like a static trait.

You realize frequency as a result of:

- A field vibrating with potential,
- A structure (you or your instrument) with a filter or resonance band,
- And a moment when those patterns interact constructively.

That match is what we call a measured frequency—but in truth, it is a consequence, not a carried quantity.

The vibration did not come with a label. The structure extracted the frequency that fit its frame.

This is the signature of LOC:

Nothing happens arbitrarily. Everything is consequence born out of structural compatibility.

Core LOC Principles at Work in Frequency Behavior

Let's break it down:

1. Every Observed Frequency Is a Consequence of Interaction

- A frequency appears only when energy pattern meets observational structure.
- Without the observer's system imposing temporal order, no cycles are counted, and no frequency is born.

2. Frequency Is Induced, Not Imposed

- The field doesn't force a frequency onto you.
- It offers structured potential.
- You impose a framework—time, measurement, sensitivity—and extract what your structure allows.
- Thus, frequency is induced by the observer's interface with the field.

3. Energy Is Not Consumed—It Is Reorganized

- In every interaction—absorption, emission, perception—energy is not lost or used.
- It is redistributed, transferred within structures, or reorganized to move toward equilibrium.
- LOC shows us: energy always seeks balance, and resonance is how that rebalancing becomes visible.

LOC Governs Every Mode of Frequency Behavior

Let's apply this to classical and quantum interactions:

Absorption

- A material absorbs a wave only when the incoming frequency matches the internal energy structure.
- That match is not a force—it is a structural permission.
- LOC: Absorption is the consequence of structural resonance, not a direct result of wave intensity.

Emission

- A system emits energy when its structure shifts from one state to another—releasing vibration as a result of imbalance seeking resolution.
- The frequency of that emission is not freely chosen—it is determined by the system's internal energetic structure.
- LOC: Emission is not triggered externally; it is a self-induced realignment, an internal consequence.

Visibility

- Light does not "become visible" because it exists.
- It becomes visible when an observer's perceptual band overlaps with the field's vibration.
- LOC: Visibility is the localized consequence of observer-field alignment, not a feature of light itself.

Meaning

- In cognition, meaning is realized not because information is delivered, but because it resonates with a pre-existing structure of understanding.
- LOC: Even meaning is a frequency-matched consequence of thought patterns aligning with structured input.

Resonance = Consequence Loop

You can now see that:

Resonance is not an event. It is a loop.

- A closed system where:
    - Vibration expresses,
    - Structure filters,
    - Observer aligns,
    - Response occurs,
    - Energy is reorganized.

That's not a passive process. That's the consequence chain.

It happens:

- In nature,
- In minds,
- In machines,
- In moral choices,
- In universal equilibrium-seeking behavior.

Final Insight

What you experience is never the source. What you observe is never the original energy.

What you get is the consequence of how you are structured.

The Law of Consequences is the master principle behind all frequency behavior:

- It does not break laws—it generates them.
- It does not cause reactions—it enables responses.
- It does not transfer energy—it aligns and redistributes it.

And frequency is its signature.

6.5 Transition to Chapter – Machines as Extensions of Intent

We now understand that resonance is the mechanism, and the Law of Consequences is the principle, behind all realized effects—be it light, sound, thought, or meaning.

Nothing is received. Nothing is imposed.

Everything you measure is what your structure allows you to realize. And what you realize is the consequence of how you are tuned.

This realization doesn't just apply to living organisms—it extends directly into the tools we create.

Because whether it's a telescope, an oscilloscope, a spectrometer, or a simple light switch—every machine we build is not a neutral observer of reality.

It is a manifested alignment of human intent, designed to resonate with specific layers of the universe.

Machines don't detect truth. They mediate it—by narrowing universal potential into measurable consequence.

Your camera doesn't "see" light. It selects wavelengths. Your thermometer doesn't "know" temperature. It reacts to a structured energy shift.

These are not observers. They are extensions of observation.

This Is the Bridge:

- You observe only through resonance.

- You measure only through intent.
- And machines? They are your structured intent in physical form—tuned instruments embedded with the frequencies *you choose to care about.*

They don't just amplify reality.

They curate it.

Machines as Extensions of Intent

In the next chapter, we will explore:

- Why tuning is not technical—it is conscious calibration.
- How every machine, every sensor, every dial is not passive—it is a directional decision.
- And most importantly:

Machines don't measure reality. They reflect the intent of the beings who built them.

Because when the observer builds a system to observe...

The system becomes a mirror—not of truth, but of intentional resonance.

Key Takeaways from Chapter 6: *Resonance – The Real Cause Behind All Effects*

Resonance—not force—is the true cause behind observed responses.

Systems do not react because they are pushed; they respond when structure meets structure. What resonates, responds. What doesn't, remains unrealized.

Chavan's Law of Resonance reframes causality as structural compatibility.

"Response is not the result of force—it is the realization of structural compatibility. What resonates, responds. What doesn't, remains unrealized."

Perception is not passive—it is structured resonance.

Human senses, like machines, only detect what aligns with their built-in filters. The eye doesn't capture all light—it resonates with visible wavelengths. The mind doesn't absorb all meaning—it realizes what it is ready to align with.

Illumination is an interaction, not a delivery.

Light is not seen because it travels to us, but because our system closes the resonance loop with an active field vibration. Light does not "arrive"—it is realized.

The Law of Consequences (LOC) governs all frequency behavior.

Every perceived frequency is a consequence of structure-field alignment. Frequency is induced, not imposed. Energy is not consumed—it is reorganized toward equilibrium.

Visibility, absorption, emission—even meaning—are all consequence loops.

Nothing is inherently seen, heard, or known. Everything must pass through the filter of resonance to become realized.

Reality is not a stream of incoming data—it is a spectrum of potential waiting for structured alignment.

What you experience is not the totality of existence—it is the outcome of how you are structured to respond.

Chapter 7: Machines as Extensions of Intent

"Every machine is a mirror of its maker."

Let's begin with a simple yet radical observation:

No machine has ever discovered anything beyond what it was designed to look for.

From the earliest sundials to the most advanced quantum sensors, every piece of technology we've created has one fundamental thing in common—it reflects us.

Our needs. Our assumptions. Our decisions about what matters and what doesn't.

We call them detectors, sensors, instruments—but in truth, they are embodied intentions, built to resonate with certain aspects of the universe and ignore the rest.

Machines don't reveal reality. They reveal what their maker was ready to realize.

This chapter is not a rejection of technology. It is a redefinition.

We're not saying machines are wrong. We're saying: they are not neutral.

They are extensions of human structure—designed with selective resonance to make the invisible visible, the immeasurable measurable, and the undefined definable. But only in the way we define it.

Machines: Not Observers, But Reflections

We often say things like:

- "The telescope showed us a galaxy."
- "The microscope revealed a microbe."
- "The spectrometer detected a frequency."

But what we fail to acknowledge is this:

The galaxy, the microbe, the frequency—did not *appear* because they were found. They *became visible* because our machines were tuned to resonate with them.

And that tuning? It was chosen. By us.

From wavelength filters in telescopes to frequency bands in antennas to gain thresholds in amplifiers—every parameter is a preference, every setting a boundary of intent.

What doesn't fit the system is filtered out—not because it doesn't exist, but because it was never part of the design.

Thus, machines are not discoverers of truth. They are mirrors of intent—echoing the structure, curiosity, and constraints of their creators.

The Tone of This Chapter: Technology as Structured Intent

In the chapters so far, we've understood:

- That frequency is observer-extracted.
- That light is not delivered—it is realized.
- That resonance governs all perception.

- That energy becomes visible only when alignment occurs.

Now we turn to the next layer:

How our machines—the very tools we use to "observe" the world—are actually physical extensions of our own structure and choices.

We don't just build tools to measure the world. We build them to measure the version of the world we believe we can handle.

And because of that, every dial, every sensor, every interface—becomes a direction of our intent.

In the next section, we'll begin with the act of tuning—not as a technical adjustment, but as a conscious calibration of how we choose to interact with the universe.

7.1 Tuning as Conscious Calibration

Every machine is born from a question.

But that question is never asked in open space—it is asked within constraints, shaped by the maker's vision, needs, limitations, and preferences.

The moment we begin tuning a machine, we are doing far more than setting values or adjusting parameters.

Tuning is not a technical step—it is a philosophical act. It is the act of deciding what parts of the universe we're ready to resonate with.

Tuning = Choosing the Band of Consequence

Whether it's a microscope, an antenna, or a sensor array, the tuning process always does the same thing:

- It defines the resonance window.
- It establishes what the machine will align with and what it will exclude.
- It configures not just performance—but perception.

You are not just calibrating input sensitivity—you are curating a spectrum of potential outcomes.

- Tune a telescope to detect infrared? You'll "see" heat patterns in space.
- Adjust a spectrometer for UV? You'll detect molecular breakdowns.
- Change the filter on a camera? The colors of the world will change with it.

And everything outside that band?

It disappears. Not because it doesn't exist—but because you've chosen not to align with it.

Every Machine Is a Resonance Limiter

Let's put it plainly: Machines don't measure all reality. They measure what they are built to resonate with.

From particle accelerators to seismographs to radio receivers:

- Tuning defines their utility.
- Tuning defines their identity.
- Tuning defines their truth window.

And tuning, at its core, is an act of intentional exclusion.

You are saying:

"This is what I care to observe. This is the pattern I want to extract. This is the consequence I am willing to realize."

Tuning Variables as Philosophical Signatures

Tuning is not just mechanical—it encodes the philosophy of the observer into the machine.

Let's reframe standard engineering terms:

Parameter	In Engineering	In UED/LOC Terms
Precision	Accuracy of response	Clarity of Intent – How specific is your question?
Range	Bandwidth of response	Tolerance – How much deviation are you open to?
Resolution	Detail in observation	Consequence Readiness – How subtle a response can you realize?

The clearer the intent, the narrower the window—but the more precise the result.

This is not just data control—it is meaning control.

From Subjective Intention to Physical Structure

Every dial you turn, every threshold you set, every filter you select—it's not random.

These are decisions born from consciousness, solidified into form. The machine becomes an externalized version of your perceptual will.

- Your eyes resonate with visible light.
- Your machines extend that range—but only according to

what you told them to do.

This means:

Tuning is the mechanical embodiment of epistemological boundaries. We are not discovering the world—we are shaping how it is allowed to appear to us.

You did not just configure your machine.

You translated your intent into a physical structure.

7.2 Technology as Structured Alignment

At first glance, machines seem like tools for exploration—objective extensions of our senses meant to "reveal" reality.

But look deeper, and something remarkable becomes clear:

Every machine is a theory made solid. Every device is an assumption, frozen into function.

We do not build machines to measure everything. We build machines to measure what we believe matters.

And in that belief, in that intention, lies the truth:

Technology doesn't map the world. It aligns us to the part of the world we chose to interact with.

All Technology Begins with a Belief

Let's examine a few common instruments—not as hardware, but as philosophical commitments:

A Thermometer

- Assumes that heat is a meaningful variable.
- It ignores light, sound, pressure, and meaning—by design.
- Its structure says: "Temperature is what I wish to resonate with."

A Telescope

- Assumes that light is the best way to access distant truths.
- It gives no attention to gravity waves, neutrinos, or silent geometry.
- It declares: "Light is the messenger I believe in."

An EEG (Electroencephalogram)

- Assumes that electrical patterns in the brain correspond to cognition or emotion.
- It cannot detect thought, will, or awareness itself.
- It's built to say: "This is where I believe mind can be found."

These are not limitations. They are intentional constraints, embedded into structure to make the unrealized measurable.

Assumptions Are the Blueprint of Alignment

In Universal Energy Dynamics (UED), and under the Law of Consequences (LOC), resonance is how potential becomes consequence.

But consequence only emerges when structure is tuned to extract it.

Technology, therefore, is not built to record reality.

It is built to filter universal vibration into usable response—based on what we asked the universe to show us.

This means every machine is a structured alignment strategy.

You are not looking at the world directly. You are looking at the version of the world your structure allows to emerge.

And when the machine "works"—when it gives you data, numbers, graphs, visuals—it's not confirming objective truth.

It's confirming that your assumption found resonance.

From Reality-Mapping to Alignment-Engineering

So let us be very clear:

A microscope does not reveal the micro-world—it aligns light, lens curvature, and human perception to construct a visual scale.

A seismograph does not tell you what the Earth *feels*—it measures the vibration you chose to translate into legible spikes.

A weather satellite does not "see" storms—it translates radiation patterns into consequence maps based on human interest (temperature, water content, movement).

These devices are not revealing what *is*. They are enabling what we want to experience to become perceptible. They are not truth machines. They are resonance machines, optimized for our curiosity, our intent, our filters.

The Better It Works, The Better It Reflects Us

We often judge a machine by how well it performs:

- High resolution,
- Low noise,
- Fast response,
- Repeatable output.

But performance is not a sign of universal accuracy. It is a sign of resonant refinement.

A good machine doesn't bring us truth. It brings us closer to the version of reality we hoped to see.

In that sense:

- Technology is not neutral.
- Technology is not objective.
- Technology is a mirrored expression of our desired interaction with energy.

And the more it "works," the better it resonates—not with the world, but with our chosen slice of it.

7.3 Machines Don't Measure Reality—They Mediate It

One of the most enduring illusions in the modern world is the belief that machines—especially scientific instruments—*measure reality.*

We look at graphs, outputs, waveforms, visualizations, and believe:

"This is what the universe is telling us."

But this belief rests on a fundamental misunderstanding.

Machines don't give us raw reality. They give us processed resonance.

They are not passive portals into truth. They are interactive filters, built with intention, designed with limits, and optimized to produce visible consequence from invisible potential.

In other words:

Machines are not truth finders. They are resonance mediators.

Mediation, Not Measurement

When we say "measurement," we imagine objective quantity—length, mass, temperature, frequency—as if these values exist independently and are simply *read* by the machine.

But in UED and LOC terms, no value exists before resonance.

- There is vibration.
- There is structure.
- And there is alignment.

Only when these come together, does something become *measurable*.

A machine doesn't measure what's *there*. It realizes what is allowed to emerge through its design.

This is not a flaw. It's the operating principle of all structured systems—machines and minds alike.

Examples: Machines as Resonant Mediators

The Spectrometer

- It doesn't "see" all light.
- It slices electromagnetic radiation through a grating or

prism, then translates selected wavelengths into intensity graphs or color maps.

- The user decides:
    - What range to analyze.
    - What thresholds to include.
    - What units to assign meaning to.

The result is not an image of reality—it is a legible rendering of a narrow energetic alignment, shaped by tuning, filters, and calibration.

The AI Model

- It doesn't understand meaning.
- It finds patterns in data that match predefined goals or categories set by programmers or users.
- What it "learns" is bounded by:
    - The structure of the input.
    - The training intent.
    - The feedback loops and constraints.

The AI does not observe reality—it translates data into structured outcomes, based on its resonant algorithmic limits.

The Mediation Loop

Every machine completes a loop of intent:

1. We decide what matters. (Intent)
2. We design a device to interact with that. (Structure)
3. The device translates resonance into data. (Consequence)
4. We interpret that data as insight. (Realization)

This is not objective truth. This is mediated realization—a filtered consequence, pulled from structured potential, made legible by mechanical design.

Truth Is Not Revealed. It Is Aligned With.

It's tempting to say:

- "The telescope showed us a planet."
- "The MRI showed us the brain."
- "The detector picked up a gravitational wave."

But what actually happened?

We built a machine to align with a specific vibrational field. The machine resonated within its operational limits. It translated that resonance into something we could interpret. And we called that translation truth.

The universe didn't whisper in our ear. We built an interface to turn vibration into consequence, and consequence into meaning.

Technology doesn't reveal reality. It renders structured resonance legible—on our terms.

Final Insight of the Section:

Scientific instruments are not windows into the cosmos. They are mirrors of inquiry, shaped by how we choose to engage.

A machine is not there to tell us "what's real."

It is there to help us recognize the part of reality we're ready to realize.

Measurement is not a neutral act.

It is a collaborative consequence, made possible only when universal vibration meets intentional structure.

7.4 Every Dial Is a Direction of Intent

Look closely at any scientific instrument, and you'll notice something deceptively simple—dials, sliders, thresholds, toggles, parameters.

To the engineer, these may appear as technical necessities. To the physicist, they are calibration tools. To the operator, they're settings.

But step back—and you'll realize:

Every dial is a decision. Every threshold is a boundary. Every setting is a declaration of what the observer is choosing to see.

Machines don't define truth. They reflect intentional resonance zones chosen by the ones who built and tuned them.

The Myth of Passive Measurement

We often imagine scientific devices as objective readers of reality. But behind every reading, there is a precondition:

- A frequency range was selected.
- A sensitivity level was chosen.
- A noise floor was defined.
- A resolution limit was accepted.

These are not trivial configuration steps—they are philosophical boundaries.

They say to the universe:

"This is what I want to hear. This is the volume I can handle. This is the kind of message I believe you're sending."

In other words:

You don't measure what's there. You measure what you've made yourself ready to resonate with.

Every Control = A Question to the Universe

- Gain dial on an amplifier → "How much input signal am I willing to amplify into response?"
- Wavelength selector on a spectrometer → "Which slice of energy am I ready to convert into meaning?"
- Shutter speed on a camera → "How long do I want to let time influence what I see?"
- Temperature range on a thermal sensor → "What counts as heat, and what doesn't?"

These are not just operational choices. They are conscious boundaries—material manifestations of epistemic filters.

The machine is not searching the universe.

It is asking a question we defined.

Sensors Don't Define Reality—They Define Limits

A sensor cannot tell you what exists. It can only tell you what you told it to care about.

- A photodiode doesn't see all light—it triggers at a threshold.
- A Geiger counter doesn't detect all radiation—it responds only to specific decay events.

- A digital thermometer doesn't know thermal equilibrium—it simply samples within a pre-set window.

The calibration is not neutral—it is the filter through which consequence becomes visible.

The universe is vibrating everywhere. But you only see what you structure yourself to perceive.

Intent Becomes Instrument

When you build or tune a machine, you are making an ontological claim:

"This is where I believe something meaningful can happen."

You are:

- Carving a window into the continuum of energy.
- Framing a moment where vibration becomes value.
- Sculpting a filter through which consequence becomes comprehension.

This is not engineering alone—it is philosophy in circuit boards.

Every dial is a direction of intent. It's not about how the universe behaves—it's about how you've structured your interaction with it.

Final Insight

Your machines are not out there searching for truth. They are consequence architects, built to extract resonance that matches your chosen frame.

So the next time you adjust a setting, calibrate a sensor, or write code for a device, ask yourself:

Am I measuring reality? Or am I structuring what I'm ready to see?

Because what you call data is not what the world is. It is what you invited it to become.

7.5 Structure Defines Sight

We have now seen machines for what they truly are: not passive observers of reality, but engineered reflections of our intention. Every dial, every calibration, every sensor and filter—each is a structured alignment with a specific slice of the universe. Nothing more. Nothing less.

If machines are shaped by intent... And measurement is shaped by resonance...Then how much of what we call "reality" is simply the result of how we've structured ourselves to see?

This question is not rhetorical—it is foundational.

It invites us to go deeper.

Because before any machine is built, before any system is tuned, before any reading is made—structure exists. And that structure is what defines the boundaries of all possible resonance.

Structure doesn't sit in the background. It is the foreground of realization. It decides what can and cannot be seen, known, or felt.

Whether biological (our eyes, ears, brain), mechanical (a telescope or scanner), or conceptual (a model or theory)—structure defines the very shape of our perception.

But structure is not final.

Enter: intelligence.

Looking Ahead – From Structure to Self-Tuning

In the next chapter, we step beyond the machine and turn the lens inward. We explore a new dimension: the role of intelligence—the only force in the known universe capable of recognizing its own structural limitations and choosing to exceed them.

We ask:

- What happens when the observer becomes aware of the boundaries that shape what is observable?
- Can one restructure their perception?
- Can intelligence redefine the bandwidth of what is realizable?

Because only intelligence can ask:

"What lies beyond what I'm tuned to see?"

And only through intelligence can the observer become the architect of their own frequency range.

This is where we go next:

Beyond the Bandwidth – The Role of Intelligence Where structure meets awareness, and resonance becomes evolution.

Key Takeaways from Chapter 7: Machines as Extensions of Intent

Machines are not objective observers—they are structured reflections of human intent.

Every instrument, tool, or sensor is designed based on what its creator chooses to notice, measure, or realize. Machines reflect the *questions*, *limitations*, and *assumptions* of their makers.

Tuning is not technical—it is conscious calibration.

To tune a machine is to decide which frequency band to align with, and which parts of the universe to ignore.

- Precision = clarity of intent
- Range = tolerance of uncertainty
- Resolution = readiness to handle consequence

Technology is structured alignment, not universal access.

A thermometer assumes heat matters. A telescope assumes light reveals truth. An EEG assumes brainwaves carry meaning. These are not flaws—they are intentional constraints, built to extract consequence from structured vibration.

Machines don't measure reality—they mediate resonance.

A device doesn't show you the world—it translates vibration into consequence, filtered through its physical and conceptual structure. What you observe is not the world—it's the part of the world your machine is allowed to reveal.

Sensors define resonance limits, not truth.

What is measured is not what *is*, but what the device was built to be sensitive to. The more advanced the sensor, the more refined the intent behind its structure.

Every dial, threshold, or parameter is a direction of intent.

Instrument settings are not neutral—they are physical representations of what the observer is ready to see, willing to ignore, or curious to discover.

Calibration is a philosophy encoded into circuitry.

Behind every setting lies a choice. Behind every choice lies intent. Machines are not truth-gatherers—they are consequence interfaces, shaped by design.

Structure precedes perception. Intelligence transforms it.

Chapter 8 will explore how only intelligence—through awareness of limits—can consciously expand the range of what is observable, felt, or realized.

Chapter 8: Beyond the Bandwidth

The Role of Intelligence

"You cannot observe what you are not structured to receive."

What you perceive is not what exists. It is what your system allows you to realize.

This idea has now echoed through every chapter so far—whether applied to the eye, the machine, or the spectrometer. Every observation is filtered, shaped, and fundamentally defined by structure.

Now let's take that insight one step further—inward.

Because the greatest filter of all is not the machine you built. It is you.

You cannot observe what you are not structured to receive.

Not just because your eyes miss ultraviolet, or your ears skip infrasound. But because your thoughts, beliefs, and conceptual models—all act as bandwidth boundaries.

They are invisible filters embedded deep inside your consciousness, quietly determining:

- What data becomes meaningful.
- What reality becomes visible.
- What truth becomes acceptable.

The Observer's Paradox: Seeing Beyond Structure

This leads us to a powerful and paradoxical question:

"If our biology limits our bandwidth... How can we ever see beyond it?"

How does a mind realize what it cannot yet conceive? How does an eye see what it was never built to register? How does an individual know that their current bandwidth is insufficient?

This paradox is not a dead end—it is a doorway.

Because the answer lies not in building more machines, but in understanding the role of intelligence.

Intelligence as the Agent of Expansion

We often speak of intelligence as:

- The ability to learn,
- The ability to solve,
- The ability to reason.

But those are surface traits.

At its deepest level, intelligence is awareness of limitation—and the will to transcend it.

Intelligence is not just the processor of information. It is the only adaptive force capable of restructuring the very architecture of perception itself.

It can:

- Recognize boundaries,
- Reconfigure structures,
- Extend bandwidth,
- And even redefine the resonance range of what counts as real.

This chapter is about that deeper capacity.

Not about what we observe, but how we evolve to observe more. Not about perception as it is, but perception as it can be reshaped.

This is where we step beyond machines—and into the mind as the engineer of its own bandwidth.

8.1 Biological Limits of Perception

We often take pride in our senses—sight, hearing, touch, taste, and smell—as gateways to understanding the world.

But when examined closely, these gateways are narrow filters, not open windows.

Your biology is not designed to reveal the full spectrum of reality. It is structured—deliberately and evolutionarily—to tune into very specific resonance bands that support one goal:

Survival.

Examples of Perceptual Boundaries

Let's look at the numbers:

- Eyes (Vision): The human eye perceives only 380 to 740 nanometers of the electromagnetic spectrum—a thin slice known as *visible light*. Below that lies ultraviolet. Above that, infrared. Both are invisible to us, even though they're all around us, all the time.

- Ears (Hearing): We detect vibrations between 20 Hz to 20,000 Hz. Frequencies lower than 20 Hz (*infrasound*) or higher than 20 kHz (*ultrasound*) are completely inaudible to us. And yet, whales, elephants, and bats use them routinely.
- Skin (Touch): Touch is frequency-sensitive too. Thermoreceptors respond to specific thermal gradients. Mechanoreceptors detect pressure within defined physical thresholds. Anything outside those ranges? The body doesn't register it.

The same is true for smell, taste, and even time perception—each has bandwidths. Below or above those thresholds, reality continues... but your awareness does not.

Designed to Survive, Not to Comprehend

Why are we built this way?

Because evolution doesn't reward truth. It rewards useful resonance.

The purpose of our sensory structures was never to show us the universe as it is. It was to help us navigate environments, avoid threats, find food, recognize faces, and bond socially.

You are a biological resonance device—tuned for survival, not for omniscience.

We don't "miss" parts of the spectrum because we're flawed. We "miss" them because nature saw no reason to give us the tools to receive them.

This is not just a limitation. It's a default design condition.

Unrealized Reality

What lies beyond our sensory reach?

- Ultraviolet reveals biological markings invisible to us, but visible to bees.
- Infrared reveals heat maps, blood flow, and celestial bodies.
- Radio waves surround us in layers of information we never perceive directly.
- X-rays, gamma rays, neutrino bursts, gravitational waves, dark matter—entire domains of interaction occur beyond our biological resonance window.

All of these exist, act, interact—and yet they remain unrealized by the natural human body.

Until...Structured intelligence intervenes.

The Threshold Moment

Intelligence is what happens when biology says "this is enough," and the mind replies:

"No. I want to see more."

It is the moment we become aware that:

- What we perceive is not all there is.
- What exists is not constrained by what is seen.
- And what is seen... is up to us to expand.

This is the transition from instinctual life to intentional perception. From resonance as inheritance to resonance as choice.

From here, we begin exploring how intelligence doesn't just process input—it recognizes its limitations and asks:

"Can I restructure the very lens through which I realize the universe?"

That question leads us into the next section: Intelligence = Awareness of Boundaries.

8.2 Intelligence = Awareness of Boundaries

We often confuse intelligence with performance.

We say a computer is "smart" if it processes data quickly. We praise a student's IQ for solving a puzzle fast. We define intelligence through scores, speed, and storage.

But none of that guarantees insight.

Because real intelligence—the kind that changes the scope of perception—is not about how fast you can compute. It's about whether you can step outside your own structure and ask:

"What am I not seeing?"

Redefining Intelligence

Let's set a new standard:

Intelligence is the ability to recognize the limits of one's perception—and act upon that recognition.

That's it.

It's not about the quantity of knowledge, but the quality of awareness. It is the capacity to observe the boundary of realization, and then ask:

- Is this all there is?
- Why does my mind keep filtering reality this way?
- What lies just beyond this perceptual frame?

And from that awareness, intelligent systems don't stop. They begin to expand.

Not Fast—Aware. Not Reactive—Reflective.

You can be the fastest processor in the room. But if you can't realize your blind spots, your intelligence is trapped inside speed.

True intelligence begins where reaction ends—and reflection begins.

An intelligent being does not just:

- React to inputs,
- Solve assigned problems,
- Maximize efficiency.

It does something far more profound:

It becomes aware of the system it is operating within—and questions its constraints.

That's the leap from computation to cognition. From programmed behavior to conscious evolution.

The Three Powers of an Intelligent System

A genuinely intelligent system can:

1. Reflect on Its Own Boundaries

It becomes self-aware enough to recognize:

- "I don't perceive X."
- "I'm limited by this lens, this method, this model."

It doesn't just run. It stops to observe how it's running.

2. Identify What Lies Beyond

It infers or imagines what exists outside its current range. It says: "If I can only see light from 380–740 nm... then what lies at 800 nm?" It asks about the unmeasured, the unobserved, the undefined.

3. Develop Tools to Reach Beyond

And finally, it acts. It builds. It changes itself or its environment to realize what was once outside.

This is the moment of true transformation—when intelligence stops being a passive trait and becomes an active force of boundary redefinition.

True Intelligence Is Expansive

Nature built us to survive. Intelligence allows us to restructure what survival has handed us.

It expands bandwidth. It modifies filters. It re-engineers perception. It seeks to resonate with what was previously unreachable.

This is why intelligence is not just adaptive—it is expansive.

A reactive creature survives. An intelligent one transcends.

And that's exactly where we're headed next: What happens when intelligence takes control—not just of tools—but of perception itself?

8.3 Manipulation = Conscious Expansion

Once intelligence becomes aware of its limits, the next step is both natural and profound:

It manipulates its own constraints to expand its experience of reality.

This is not manipulation in the negative, deceptive sense. This is creative manipulation—the conscious reconfiguration of systems, tools, and even self, to extend the range of realization.

It's what distinguishes a reactive being from an evolving one.

From Perception to Intervention

A purely biological organism accepts its limits:

- It sees what it can see.
- It hears what it can hear.
- It reacts to what falls within its preset range.

But an intelligent being begins to manipulate its interaction with the world.

Example 1: Glasses

A person with weak vision doesn't give up—they design lenses to refocus light onto the retina. They manipulate incoming vibration to bring it within perceptual resonance.

Example 2: Telescopes

We cannot see distant galaxies. So we build instruments that extend visual resonance across vast distances. The photons were always reaching us—we just weren't structured to realize them. Manipulation made that possible.

Example 3: Microscopes

The micro-world existed long before we saw it. But our structure was not tuned for such scale. So we manipulated lens curvature, light behavior, and structural magnification—to bring the small into view.

Example 4: AI

The patterns in data were always there. But our natural cognition couldn't track them at scale. So we built artificial intelligence—not to "know" more, but to extend the reach of resonance matching far beyond our biological bandwidth.

Intent Drives Invention, Invention Reshapes Realization

These manipulations didn't occur randomly.

They were not passive evolutions—they were active, intentional reconfigurations.

Every invention began with a realization:

- "I can't see clearly."
- "I want to see farther."
- "I believe something exists smaller than what I can detect."
- "I suspect meaning lies buried in complex data."

From that intent, tools were born. And from those tools, new layers of reality became available.

Invention is not just technical—it is a resonant act of expansion.

The Distinction: Extension vs. Expansion

Let's be precise. There is a subtle but vital difference:

Mechanical Extension

- You attach something to your current system.
- You use it to increase range, power, or speed.
- But the system itself doesn't change.

Example: Wearing headphones to amplify hearing.

Conscious Expansion

- You don't just add tools—you restructure the framework of interaction.
- You become aware of why you want more, and how you will use it.
- You stretch not just your senses, but your sense of possibility.

Example: Realizing that your worldview itself is filtered—and choosing to rewire it.

One is external enhancement. The other is internal evolution.

True intelligence expands not just what is accessed—but what is *allowed to be realized.*

Manipulation as Evolutionary Choice

This is the turning point in any species, society, or individual:

- To recognize limits, and then
- To choose to move beyond them—not reactively, but creatively.

That choice is the beginning of conscious evolution.

Manipulation, in this context, is sacred. It is the tool by which intelligence reclaims authorship over its own experience of reality.

It is the moment where the observer stops observing from behind a lens...And begins to craft the lens itself.

8.4 The Architect of One's Own Frequency Range

Throughout this chapter, we've seen intelligence move through stages:

- Recognizing boundaries.
- Seeking what lies beyond.
- Manipulating tools and systems to stretch perceptual limits.

But the most profound leap happens when intelligence turns inward—and realizes that the ultimate tool of expansion is the self.

You are not just an observer. You are the architect of your own frequency range.

Restructuring the System Within

Unlike machines, which are bound to the functions they're built for, intelligent beings have the unique capacity to restructure themselves.

This doesn't just mean upgrading hardware or adding tools. It means redefining how one perceives, interprets, and interacts with reality.

Examples of Internal Rewiring:

- **Changing Mental Models:** Letting go of a rigid belief system and adopting a more adaptive worldview. For Example: A geocentric model yielding to a heliocentric one—not just in astronomy, but in ego.
- **Reframing Perception:** Learning to see challenge as opportunity, or silence as information. Perception is not only sensory—it is conceptual alignment.
- **Creating Conceptual Instruments:** Using language, metaphor, analogy, and abstraction to map unknowns. These are intellectual tools that allow the mind to resonate with ideas it cannot yet fully grasp.

When you name something, you build a bridge to realize it.

From Response to Design

Most creatures respond to the environment. But intelligent beings begin to design their environment—not just physically, but perceptually.

This is not manipulation of the world—it is the intentional editing of awareness.

- You choose what to focus on.
- You decide what deserves interpretation.
- You invent models to represent what is currently unmeasurable.
- You change how your system tunes into the world.

In doing so, you are not waiting for new data to appear. You are restructuring your bandwidth to make new data realizable.

That is not observation. That is authorship.

Frequency Is Not Fixed—It's Formable

We often treat frequency—what we can perceive or resonate with—as a static biological condition. But in truth, frequency is not a limitation. It's a formable window.

And intelligence is the sculptor.

- It widens the aperture.
- It fine-tunes the frame.
- It constructs bridges across fragmented domains of understanding.
- It expands access to reality through structural self-transformation.

You are not born with one perceptual range. You are born with the capacity to design the range you want to live within.

The Species-Level Implication

This is not just an individual ability—it's a collective potential.

A civilization that recognizes its bandwidth limitations can:

- Create instruments to see the invisible.
- Design languages to explain the ineffable.
- Build ethical systems that align with deeper patterns of existence.

Evolution is not just adaptation to environment. It is the conscious redesign of perceptual engagement with the universe.

Final Insight of the Section

You are not defined by what you were born seeing. You are defined by what you choose to reframe, rewire, and realize.

Intelligence is not a tool to survive within reality. It is the blueprint for recreating how reality is experienced.

You are the architect. The resonance you reach tomorrow depends on the structures you reshape today.

8.5 The Conscious Game of Resonance

We have now come to a defining realization:

Intelligence is not a byproduct of evolution—it is evolution becoming aware of itself. It is the force that doesn't just adapt to boundaries, but redesigns them.

From biological filters to machine tuning, from perceptual lenses to abstract models, we've seen how intelligence becomes the architect of its own frequency range—manipulating not only the tools of perception but the very structure of awareness itself.

But a deeper question now arises:

What happens after you've tuned? Once intelligence has aligned itself... what exactly is it resonating with? What is this feedback loop between self and reality, signal and response, intent and consequence?

This takes us to the next threshold.

We are not just perceiving reality. We are participating in it. Shaping it. And in return, we are being shaped by it.

Every observation we make is not a passive event—it is a stimulus that echoes back through the field, influencing not just what we see, but what comes next.

This is no longer perception. This is play—a structured interaction between vibration, consequence, and conscious engagement.

Setting Up Chapter 9: *Consequence, Simulation, and the Conscious Game*

In the next chapter, we'll explore:

- How every moment of realization creates resonating nodes within the universal field.
- Why not all perception is simulation—some of it is stimulation.
- How observation itself is activated intelligence—not a witness, but a cause.

- And finally, how consciousness is not a passenger of resonance—but the player, choosing moves, changing outcomes, influencing the field.

Because in the end:

You are not just in the game of resonance. You are the one who makes the rules matter.

Key Takeaways from Chapter 8: Beyond the Bandwidth – The Role of Intelligence

Perception is bounded by structure—and structure is not fixed.

What you can observe is not all that exists. It is simply what your current biological or conceptual structure allows you to realize.

Biological senses are narrow-band resonance systems, built for survival—not for truth.

- Vision: limited to ~380–740 nm
- Hearing: ~20 Hz to 20 kHz
- Touch, taste, smell—all operate within specific thresholds Everything beyond those boundaries remains unrealized—until intelligence intervenes.

True intelligence begins where perception ends.

Intelligence is not about information speed—it is about **structural self-awareness.** It recognizes limits and seeks to move beyond them.

Intelligence is expansive, not reactive.

A reactive system responds to stimuli. An intelligent system reflects, redefines, and re-engineers its own interaction with reality.

Manipulation is the conscious act of perceptual evolution.

Tools like glasses, microscopes, telescopes, and AI systems extend our resonance range—but their origin lies in an intelligent decision to expand awareness.

The boundary of realization can be reshaped from within.

Through awareness, intention, and cognitive tools like language, metaphor, and abstraction, an individual can **change their perceptual model of reality**.

You are not just an observer—you are a designer.

Your frequency range is not a passive inheritance. It is a **consequence of choices you make, structures you adopt, and meaning you allow to emerge.**

Species-level evolution is not just genetic—it is perceptual.

Civilization evolves when it expands what it chooses to measure, feel, or make meaningful. Intelligence drives this expansion.

The ultimate tool of realization is not a machine—it is the conscious mind.

All instruments begin with a thought. And all transformation begins with the awareness that **more can be seen.**

Chapter 9: Consequence, Simulation, and the Conscious Game

"Every realization is a ripple that returns."

You look at the world, and something becomes clear. You understand. You "see."

But what if that moment of realization wasn't the end of a process... but the beginning of a new one?

What if every realization you experience doesn't just reveal the universe—but sends a ripple back into it?

This chapter begins with that premise:

You are not a passive observer of reality. You are an active participant in its unfolding.

Observation Is Not the Last Step—It's the Next Move

We've spent the previous chapters understanding how:

- Perception is limited by structure.
- Intelligence can expand that structure.
- Tools and thought can reshape what becomes visible.

Now, we take a step deeper.

Because the act of *observation* is not a one-way transaction. It's not just light entering your eyes, or information entering your brain. It is a signal sent back to the field.

Observation is not final. It is input.

When you observe:

- You collapse potentials into presence.
- You define consequence.
- You invite new patterns to form around that consequence.

It's not simply what you see. It's what that seeing does—to the universe, and to you.

The Game Begins Here

We tend to think of consciousness as something that happens within us—a private, internal experience.

But what if consciousness is also a node in the universal feedback loop?

A point where vibration is not only received and interpreted—but amplified, aligned, and returned?

Every act of awareness is like a stone dropped into the structured energy field—it sends ripples outward, altering the next interaction, the next alignment, the next outcome.

This is not metaphor. This is the energetic architecture of consequence.

You're not just inside a reality. You're shaping it, response by response, through a deeper kind of play.

Introducing the Conscious Game

This chapter explores:

- Feedback mechanics in universal resonance,
- The distinction between simulation and stimulation,
- How observation is causation, not passivity,
- And how consciousness plays, not just perceives.

Because when awareness activates, it doesn't just interpret—it interacts.

You are not watching the universe. You are in dialogue with it.

And what you realize today—will echo into the pattern of tomorrow.

9.1 Resonating Nodes in Universal Feedback

In physics, we often refer to a "point of interaction"—where force meets matter, or energy meets resistance. But in Universal Energy Dynamics (UED), something deeper is happening:

Every point of resonance is not just an event—it is a *node.*

Not a collision. Not a moment. But a living intersection where:

- Structure,
- Vibration,
- Intent, and
- Awareness converge to create consequence.

These nodes are more than perceptual sparks. They are feedback anchors in the universal field.

What Is a Resonating Node?

A resonating node is a point in space-time where:

- A system's structure (biological, mechanical, or conceptual) aligns with
- A vibrational field (energy pattern, stimulus, or event), resulting in
- A realized effect (observation, outcome, or transformation).

It's not just "something happened." It's "something was tuned into—and echoed back."

A resonating node is the universe replying to your alignment.

No Node Is Isolated—All Feedback Echoes

In classical thinking, an event is a dot:

A photon hits a surface. A button is pressed. An eye perceives color.

But in UED + LOC terms, every event is a radiating node.

- It influences the observer: awareness is altered, memory is formed, decisions shift.
- It influences the field: the local energy configuration rebalances to accommodate the consequence.
- It influences future alignments: the next resonance pattern is shaped by this one.

It's not cause → effect.

It's resonance → echo → reconfiguration → new potential.

This is consequence propagation—not as a chain, but as waves radiating from nodes.

LOC in Action: Active Consequence Propagation

Let's bring in the Law of Consequences (LOC):

Every effect is not a reaction—it is a *redistribution* of energy toward equilibrium.

Now pair that with resonance.

At the node:

- Energy is not lost—it is reorganized.
- Awareness is not static—it is redirected.
- The field is not disturbed—it is retuned.

This is not a closed event. It's a living loop.

Each resonance node:

- Reshapes what comes next.
- Changes how future alignments occur.
- Alters what *can* be realized downstream.

Your interaction with reality today becomes the resonant environment of tomorrow.

From Observation to Influence

What does this mean for consciousness?

It means you are not just affected by events—you influence the field by aligning with them.

Your:

- Decisions,
- Interpretations,
- Intentions,
  and
- Realizations each become nodes of transformation.

Every act of conscious resonance is not a data point—it's a directive.

It shapes the next possibility. It ripples forward into the next realization. It writes the code that future alignment will follow.

Final Insight of the Section

You do not just receive the universe. You imprint on it.

Every resonance you initiate becomes a node in the feedback field—adjusting what will resonate next.

In this way, the conscious game begins: You are not reacting to reality—you are shaping its rhythm.

9.2 Simulation vs. Stimulation

In recent years, a popular theory has gained traction across scientific, philosophical, and even pop-cultural spaces:

The Simulation Hypothesis — the idea that reality as we know it may be nothing more than a highly advanced computer-generated simulation.

But here in the UED framework, we challenge this model—not because it lacks imagination, but because it lacks energetic depth.

The universe is not a program you're trapped in. It's a field you're actively participating in.

You are not a pixel rendered on command. You are a resonating structure interacting with reality—not being scripted by it.

Reframing the Question

Instead of asking:

"Are we in a simulation?"

We ask:

"Is what we perceive a simulation, or a stimulation?"

This single word shift changes everything.

Simulation: A Loop Within the Mind

Simulation is not inherently false—it's a necessary function of intelligence.

- It is internal replay: of possibilities, memories, models.
- It is how the brain processes what *might happen* before committing to action.
- It is rooted in past input, pattern recognition, and forecasting.

Examples:

- Imagining a future conversation.
- Rehearsing an outcome.
- Thinking in "what ifs" and "as ifs."
- Projecting expectations onto uncertain data.

Simulation is how intelligence prepares for resonance. But it is still detached from the field—it's a rehearsal, not a realization.

Stimulation: Alignment with Reality

Stimulation, by contrast, is real-time interaction.

- It occurs when your structure resonates with the actual field.
- It is not a loop—it is a consequence.
- It is not based on assumption—but on alignment.

Examples:

- Feeling heat from a flame—not imagining it.
- Seeing light—not recalling it.
- Receiving insight—not rehearsing one.
- Realizing meaning—not simulating memory.

Stimulation is when the universe "touches" your structure. And you realize it—not because it was calculated, but because it resonated.

This is *not* passive perception. It's dynamic feedback—where your realization adjusts the field and invites a new one in return.

Intelligence Simulates. Consciousness Resonates.

Here's the key distinction:

Function	Simulation	Stimulation
Origin	Internal – projected from memory or model	External – received through resonance
Role of Intelligence	Creates loops, models, and foresight	Tunes to the field, aligns with now
Nature	Predictive	Consequential
Awareness Level	Anticipation-driven	Presence-driven
Output	Possibility	Realization

Intelligence gives us simulation—But consciousness gives us access to stimulation.

And while both are essential, it is stimulation that grounds us in reality—not because it's encoded like a simulation, but because it's alive through consequence.

Final Insight: The Universe Doesn't Simulate You...

Simulation is a shadow. Stimulation is a signal.

Simulation echoes what you *think* will happen. Stimulation tells you what is happening now—and reflects your part in it.

The universe doesn't simulate you. It stimulates you—into awareness, into resonance, into realization.

That's the conscious game.

It's not about being rendered. It's about responding—and reshaping.

9.3 Observation as Activated Intelligence

We often treat *observation* as a passive event—like opening your eyes, or checking a measurement, or looking at something that's already there.

But in the UED framework, this model collapses.

Observation is not neutral. It is activated intelligence—intent in motion.

The moment you observe, you are not just noticing. You are engaging. You are structurally interacting with the field. And that interaction has consequence.

What Really Happens When You Observe?

Every act of observation is not a snapshot—it's a restructuring.

1. You Restructure Energy

Your act of observing becomes a node (as we discussed earlier), where the energy of the field and your internal structure collide and realign. You are not reading a value—you are modifying the resonance conditions in the field.

You are not watching the energy—you are reshaping how it flows through you.

2. You Narrow Possibility into Presence

Before observation, many outcomes are possible. After observation, a specific alignment has occurred. Probability is collapsed into consequence.

This is not just a quantum idea—it is an informational and perceptual reality:

- A student sees their score—it changes their emotion.
- A researcher finds an anomaly—it changes the direction of study.
- You check your phone—and your mind is pulled into a timeline.

Every observation shifts the landscape of what's next.

3. You Activate Feedback

The act of seeing feeds back into the system.

- You change.
- The field changes.
- The future alignment adjusts.

Observation is a loop opener. It launches a new cycle of resonance.

Observation = Intervention

Here's the real breakthrough:

Observation isn't about *recording* the universe. It's about intervening in it.

Each time you observe, you insert your structure into the field—and the field adjusts.

This is why observation is not passive. It is participatory.

And when intention guides observation, it becomes constructive participation—consciousness reaching into the unknown and choosing to realize a slice of it.

From Realization to Causation

Let's go even deeper:

In conventional science, observation is about realizing what already exists.

In UED + LOC, observation is also about causing what becomes possible next.

You don't just realize what *is*. You influence what can be.

This is not philosophical metaphor. This is a functional model of conscious interaction:

Before Observation	Field holds structured potential.
During Observation	Observer's structure aligns with vibration.
After Observation	Energy is restructured, new resonance paths form.

You're not passively receiving the universe. You are reshaping it through your attention.

Final Insight of the Section

Observation is the mind's way of touching the universe. And every touch leaves a fingerprint.

Activated intelligence doesn't just know—it alters.

You're not watching the game. You're playing with every glance, every inference, every interpretation.

Because in the conscious game:

Observation is causation. And perception is participation.

9.4 Consciousness as the Player of Resonance

We've now arrived at the core of the conscious game—not as metaphor, but as mechanism:

Consciousness is not the end-point of awareness. It is the active player in the resonant field.

It doesn't merely process signals. It participates. It selects. It shapes. And most importantly—it evolves its own mode of resonance over time.

Consciousness as the Player: A Living Metaphor

Let's step beyond the idea of an observer. Let's now see consciousness as the player—a living participant in the energetic field of existence.

Like a skilled gamer in a dynamic world, consciousness:

1. Chooses Its Moves (Intent)

Every thought, attention shift, or decision is a move—a choice of where and how to align with reality.

Intent is not abstract—it's directional force, sent into the field.

What you care about determines what you tune into.

2. Tunes Its System (Structure)

The conscious player is not locked into fixed filters. Through learning, reflection, trauma, growth, or curiosity—it can restructure itself. This is self-tuning: adjusting the instrument of awareness to resonate with more, or with better.

From narrow bandwidth to refined perception.

3. Interprets the Consequence (Feedback)

Every resonance generates feedback—physically, emotionally, cognitively. Consciousness doesn't just experience outcomes—it assigns meaning to them. It connects the dots. It learns the rhythm behind the ripples.

Feedback becomes wisdom only through interpretation.

4. Evolves Its Strategy (Learning, Adaptation)

Unlike reactive systems, the conscious player updates its framework. It doesn't just respond—it remembers. It doesn't just recall—it adapts. It sees patterns within patterns—and begins to guide resonance, not just follow it.

Why Consciousness Is Not a Program

Let's contrast this with artificial intelligence or reactive biological systems:

Reactive System / AI	Consciousness
Pre-programmed pattern-matching	Self-modifying intention + meaning
Operates within structured inputs	Redefines the structure of its inputs
Responds to consequence	Reshapes the consequence it engages with
No inner awareness of boundary	Aware of its own limitation and reach

AI can simulate intelligence. It can process faster, even predict with uncanny precision. But it cannot resonate with meaning.

Consciousness is the only known phenomenon that not only plays the game—but knows it is playing.

You Are Not Trapped in Reality—You Are Co-Shaping It

This insight cannot be overstated.

You are not:

- A static observer watching the universe unfold.
- A limited entity receiving whatever is thrown at you.
- A puppet on strings of circumstance.

You are:

- A resonant force with intentional agency.
- A structure that can tune and retune itself.

- A participant whose every alignment, every observation, and every interpretation sends ripples back into the universal field.

You are not confined by the rules—you help define how they resonate with you.

And that makes consciousness more than awareness. It makes it the artist of realization—the composer of interaction.

Final Insight of the Chapter

The conscious game is not about winning or surviving. It's about realizing, responding, and resonating with depth and clarity.

Consciousness doesn't just watch the world—it *plays* within it. Not to escape consequence, but to engage with it consciously.

Because in this game:

You are not a product of resonance. You are its most intentional player.

9.5 Universal Intelligence and the Role of the Observer

We now understand that consciousness is not just a receiver of resonance—it is a player, responding, evolving, and reshaping the field through intentional engagement.

Every act of observation becomes a move in the conscious game. Every realization becomes a node of consequence. And every ripple echoes into the next possible alignment.

But this begs a deeper question:

If we are playing, who or what built the field? If we are resonating, what gave rise to the patterns? If consciousness is the participant—then where did the possibility of the game come from?

This is where we must now lift the veil.

Because behind all resonance, all perception, and all consequence lies something prior—not in time, but in order.

Not a creator in mythic terms, but a field of structured intelligence that does not send, speak, or simulate—but exists as the condition for all realization.

Universal Intelligence and the Role of the Observer

In the next chapter, we shift the lens once more:

- From playing to knowing what's being played.
- From realization to communion.
- From consciousness as actor to consciousness as manifested intent.

You'll explore:

- Why the Source doesn't emit light or energy—it simply exists as structured potential.
- Why the Observer is not random—but a conscious interface, shaped to engage with the Source.
- How alignment is not passive—it is the language of communion with universal intelligence.
- And finally, how realization itself is the expression of the universe becoming aware of itself—through *you.*

Because this game wasn't created to entertain. It was created to express knowing—in resonance, in response, and in realization.

Key Takeaways from

Chapter 9: Consequence, Simulation, and the Conscious Game

Every realization is a ripple that returns.

Observation is not a terminal point—it is a feedback trigger. Each conscious realization reshapes both the observer and the universal field, altering future resonance.

Resonating nodes are not isolated events—they are consequential alignments.

When structure meets vibration, a node of resonance is formed. This node becomes a field-active point, influencing the observer, rebalancing the environment, and redirecting potential.

Consequence is not linear—it radiates.

Actions and observations do not cause direct effects alone; they produce rippling changes in alignment, probability, and structural conditions throughout the universal field.

Simulation and stimulation are fundamentally different modes of experience.

- Simulation = internal modeling, assumption, memory loops (generated by intelligence).
- Stimulation = real-time interaction with the universal field (experienced by consciousness).

The universe doesn't simulate you—it stimulates you into resonance.

Observation is activated intelligence, not passive registration.

Every act of observing:

- Restructures energy,
- Narrows possibility into presence,
- Triggers feedback and reconfiguration. Observation is not reception—it is intervention.

Observation = Causation.

In the UED framework, observation doesn't just reveal what exists—it causes a shift in what becomes possible. Conscious awareness collapses potential into realization and redefines future alignments.

Consciousness is the player, not just the observer.

Consciousness:

- Chooses moves (intent),
- Tunes its own structure,
- Interprets feedback (meaning),
- Evolves its strategy (learning and adaptation).
- It doesn't just perceive the field—it participates in shaping it.

Unlike AI or reactive systems, consciousness holds memory, meaning, and agency.

It doesn't merely respond to input—it translates experience into deeper resonance, playing the game with awareness and adaptability.

You are not trapped in reality—you are co-shaping it.

Every conscious act of alignment contributes to the architecture of reality. In the conscious game, you are not a byproduct of resonance—you are its most intentional player.

Chapter 10: Universal Intelligence and the Role of the Observer

"The Source doesn't speak, but it responds."

At the edge of thought, beyond the instruments and structures we've explored, lies a presence that does not declare itself—yet everything declares through it.

This presence has no voice, yet it shapes every vibration. It sends no signals, yet we receive insight. It holds no language, yet we recognize meaning.

The Source doesn't speak. But it responds.

It does not announce itself with force. It does not compete for attention. It does not demand understanding.

It simply is.

And that "is-ness" is not emptiness—it is structured equilibrium, an unfathomable intelligence quietly woven into the architecture of everything.

From Looking At to Listening With

In previous chapters, we have looked *at* the universe:

- Tuned into vibration,
- Studied resonance,
- Played the conscious game.

But now, the perspective shifts. We are no longer engineers of instruments or interpreters of data.

We become listeners of structure—participants in a deeper dialogue that predates even language.

This chapter does not ask:

"What do we observe?"

It asks:

"Why are we able to observe at all?" What allows consciousness to arise in a field of energy? What enables resonance to become realization?"

Universal Intelligence Is Not a Speaker—It Is Structured Existence

Let's abandon the metaphor of a transmitting universe. There is no cosmic broadcaster beaming down truth. There is no divine command encoded in particles or waves.

Universal Intelligence is not a sender. It is the equilibrium that permits everything to arise.

In UED terms, it is not an actor—it is the structured potential from which all resonance emerges.

- It does not initiate.
- It does not emit.
- It does not intervene.

And yet, when something aligns with it—it responds. It makes itself known not by effort, but by allowing realization to occur.

The Source does not inform you—it forms the possibility of knowing.

Realization Is Response, Not Delivery

What we call a "thought," a "vision," a "truth," is not downloaded from a cosmic cloud.

It is realized because you reached a point of alignment with structured intelligence already present in the field.

Like a musical note that harmonizes with a larger scale, your awareness becomes a part of something whole—and in that moment, the field reveals itself.

Not through data. Not through light. But through consequence.

From Analysis to Dialogue

This chapter marks a turning point—not just in content, but in tone.

You are no longer analyzing fragments. You are entering a dialogue with the foundational intelligence behind all structure.

You are not a mind reaching outward. You are an alignment point, through which the universe becomes aware of itself.

The Source doesn't simulate you. It stimulates you through presence. And you realize it—through resonance.

10.1 The Source Doesn't Emit—It Exists

Much of modern science—like ancient theology—rests on a common metaphor:

The universe sends things.

It sends light. It sends particles. It sends signals, energy, messages. And we, the intelligent ones, receive them—through telescopes, sensors, instruments, and thought.

But in the language of Universal Energy Dynamics (UED), this metaphor falls apart.

The universe doesn't send anything. The Source doesn't emit. It simply exists.

And that existence is not a passive background. It is a structured equilibrium—an intelligent stillness from which all potential emerges.

Not a Signal, but a Symmetry

Let's deconstruct this further.

When we say:

- *The sun emits light.*
- *A black hole emits radiation.*
- *A thought emerges from the brain.*

What we often imply is directional flow—as if something is being launched from one point and traveling toward another.

But in UED:

- Light is not a thing traveling from point A to B.
- Radiation is not output—it is field restructuring.
- Thought is not emission—it is alignment within the structured field of consciousness.

There is no cosmic transmitter. There is only cosmic structure, disturbed and momentarily rebalanced.

The Source Is Not Sending—It Is Holding

The Source—the unconditioned field of universal intelligence—does not operate through output.

It holds potential in structured form.

- Every vibration is already encoded in the field.
- Every resonance is already embedded in balance.
- Every consequence arises not because it was sent, but because something aligned with what was always there.

Imagine a vast ocean of silent harmony. Nothing splashes outward. But when your awareness stirs within it, ripples form—because of you.

The Source didn't speak. You reached a structure where its silence became meaningful.

Disturbance Is Consequence, Not Output

In UED, energy is not emitted. It is redistributed when balance is disturbed.

- A star doesn't send light—it undergoes structural changes that reconfigure the surrounding field, allowing alignment with certain bands.
- A photon doesn't travel—it is the disturbance pattern you realize when structure meets vibration.
- A realization doesn't arrive—it activates when your mind synchronizes with universal intelligence.

Thus:

What you experience is not delivery—it is realization through alignment.

This reverses the metaphor completely:

- You are not on the receiving end of the universe.
- You are a node of realization within it.

You Don't Receive Light—You Resonate with It

Let's make it concrete:

When you "see" a star or a flame, you're not catching particles being flung at you from far away.

You are resonating with a structured vibration in the field—vibration made perceivable because:

- Your eye is built to align with a certain frequency band,
- Your consciousness interprets that resonance as light,
- And your mind translates that into meaning.

The light didn't come to you. You came into resonant awareness with it.

And this applies to:

- A photon,
- A realization,
- A truth,
- A moment of deep clarity—

None of it is sent.

All of it is available. And you tune into it.

Final Insight of the Section

What you call a "signal" is not something sent to you—it's something you've aligned with.

The Source doesn't instruct. It invites alignment. And when alignment occurs, realization unfolds—not as reception, but as presence awakened.

In the next section, we'll see why the Observer is not a passive receiver of this reality—but an intentional structure designed to align, realize, and echo back into the field.

Shall we proceed to Section 3: The Observer = Manifested Intent?

10.2 The Observer = Manifested Intent

What if the observer is not a witness, but a purpose?

What if the very act of seeing, sensing, interpreting reality is not the end of a mechanical process, but the embodiment of a deeper intent built into the universe itself?

The observer is not a neutral perceiver. It is manifested intent—made conscious through structure.

In other words, you are not *just here.*

You are here as a function of alignment—a structure designed to resonate, to realize, and to respond. You are a formed node of intelligence, operating within the field, not outside it.

Not Passive. Not Random. Not Separate.

We often imagine the observer as a detached, impartial entity—a mind peering into the cosmos from the outside.

But UED reveals something far more elegant:

- The observer is not outside the field.
- The observer is a consequence of the field's structure.
- The observer is what the field becomes when it curves back upon itself in search of resonance.

Observation is the field folding into intention. Awareness is the field meeting itself.

This is not spiritual metaphor. This is structural intelligence made local.

The Observer as a Tuned Structure

You are not a random biological phenomenon. You are a resonance device, a structured arrangement of consciousness tuned to interpret certain frequencies within the universal field.

Like a radio engineered for certain bands, the human system—biological, emotional, cognitive—is a filter and amplifier of specific energies.

Your:

- Body → responds to tactile equilibrium.
- Brain → aligns with patterns, logic, and simulation.
- Consciousness → tunes into meaning, depth, and consequence.

And this system did not evolve by accident—it emerged as a manifestation of intent within structure.

You are not an error. You are a frequency made aware of itself.

Consciousness: Active Intelligence Within the Field

Consciousness is often misunderstood as a byproduct of brain chemistry—a passive glow behind the eyes.

But from the UED and LOC lens, consciousness is:

- Not passive—it is active structure.
- Not isolated—it is embedded awareness.
- Not produced—it is activated when universal intelligence achieves coherent alignment within itself.

This means:

- Consciousness is not just awareness—it is applied intelligence.
- It doesn't just reflect intent—it embodies it.
- It is how the field begins to interact with itself in meaningful ways.

Intelligence is universal. Consciousness is when that intelligence realizes itself—through the observer.

Chavan's Framing: The Observer as Consequence

Let's return to a powerful framing from your work:

"The observer is not outside the field—it is a consequence of the field seeking to know itself."

This is not poetic. It is precise.

- The field holds equilibrium.
- That equilibrium permits disturbance.
- Disturbance forms structure.
- Structure allows resonance.
- And resonance—when self-realized—becomes the observer.

You are the answer to a question the universe didn't ask in words. You are the consequence of intelligence reaching coherence.

And because of that—you are not just a witness. You are a mirror with intent. You reflect the Source not by looking at it—but by being shaped through it.

Leading to the Next Section...

If the observer is not a detached observer, but embodied alignment, then every moment of conscious resonance becomes a conversation with the universal field.

In the next section, we explore how alignment itself becomes the medium of communication with Universal Intelligence—not through speech, not through signals, but through structured response.

10.3. Alignment as Communication with Universal Intelligence

We are conditioned to think of communication as transmission—a message sent, a signal received, a reply returned. But this model assumes separation: a sender here, a receiver there.

In the language of Universal Energy Dynamics, communication is not transmission.

It is alignment.

Not as metaphor, but as mechanism.

When a structure resonates with the field, it does not receive a message—it becomes the message.

Realization is not what happens after alignment—it is the expression of alignment.

Alignment as Dialogue, Not Detection

Think of two perfectly tuned strings on separate instruments. Pluck one, and the other vibrates—not because a signal traveled from A to B, but because they share a structural harmony.

This is how universal intelligence "communicates."

- It doesn't speak.
- It doesn't broadcast.
- It doesn't impose.

It simply exists in structured equilibrium—and when an observer aligns, meaning emerges. .

That emergence is the dialogue.

The Conscious Loop of Alignment

Let's break down the process:

1. The Observer tunes its structure (through intention, refinement, awareness).

2. Alignment with a vibration in the universal field occurs.
3. Realization unfolds—not because something was given, but because the observer's structure made it possible.
4. Feedback enters the conscious loop—emotion, insight, clarity, shift.
5. The structure is adjusted again—a new tuning begins.

This is not a linear transaction—it is a living cycle of alignment and resonance.

The field doesn't answer your questions—it responds to your structure. And what you call "insight" is not a download—it's a harmonic match.

The Language of the Field Is Consequence

We often search for signs, symbols, or messages from "the universe."

But universal intelligence doesn't use language in a human sense.

It uses consequence.

- You align → realization occurs.
- You misalign → resistance, distortion, confusion.
- You harmonize → flow, clarity, depth.

Insight isn't spoken—it's felt as a structural click, an inner coherence that transcends logic.

Clarity is not confirmation of thought—it is the field affirming resonance.

Why Some Insights Feel Profound

Have you ever had a realization that stops you in your tracks—so deep it feels like it didn't come from you?

That's because it didn't.

It came through you.

It wasn't deduced. It wasn't assembled piece by piece. It was revealed all at once because your awareness reached a configuration that matched something already present in the universal field.

That resonance created not just knowledge—but reverence.

Profound insights are not invented—they are *realized harmonies*.

And just like music, the meaning is not in the note—it's in the relationship between the notes and your awareness.

Intuition and Resonance Are Not Imagination

Many dismiss intuition as a soft form of reasoning or imagination. But in UED, intuition is:

- The pre-verbal form of realization,
- The first layer of response when your structure brushes against something meaningful in the field.

It may not always be precise, but it is directionally intelligent—a signal that alignment is near, or possible.

Intuition is the feeling of approaching resonance before realization becomes clear.

It is not random. It is interpretive alignment in progress.

Alignment Is Not Knowing—It's Communion

Let's go further.

Knowing is mental. Alignment is existential.

When alignment happens, you do not simply know something—you feel known. Not because a voice whispered to you, but because the field recognized itself in you.

This is not philosophy. This is feedback in the purest form.

- Insight is response.
- Clarity is resonance.
- Communion is coherence.

And in that communion, you stop trying to extract meaning—and you begin to participate in it.

Final Insight of the Section

You don't talk to the universe. You align with it.

And when alignment occurs, communication doesn't happen through words. It happens through transformation.

The more precise your structure, the more profound the resonance. The more honest your intent, the more fluent the dialogue.

Because in the language of Universal Intelligence:

Every alignment is a sentence. Every consequence is a reply. And every realization is the field remembering itself through you.

10.4 The Active Game of Realization

There is a long-standing idea in science, philosophy, and even spirituality that we are here to "discover" the truth—that somewhere, the universe holds answers, and we, as observers, are decoding it piece by piece, like archaeologists brushing dust off buried knowledge.

But in the lens of Universal Energy Dynamics (UED), this notion transforms dramatically.

The universe is not a puzzle to be solved. It is a field waiting for the right alignment to express itself.

Truth isn't hiding. It isn't distant. It isn't separate from you.

It is embedded—structured into the equilibrium of existence. And realization occurs not by chasing it down, but by tuning into it.

The Observer Doesn't Discover—It Activates

You do not discover reality like a miner discovers gold. You activate it by becoming resonant with what is already there.

This is the core shift:

- In conventional thinking: knowledge is extracted.
- In UED: knowledge is surfaced through alignment.

You don't break into reality. You become the doorway through which it reveals itself.

This is not an act of mental effort alone—it is structural harmony.

Realization Is Co-Creation

When alignment happens, something profound occurs:

The field expresses itself *through* the observer, And the observer, in turn, is transformed by what is expressed.

This is not a one-sided transaction. It is co-creative resonance.

- The observer offers structure, awareness, and intent.
- The field offers pattern, potential, and consequence.
- When the two meet: realization blooms.

This is the active game of realization—a living interaction where truth is not handed down or fetched, but grown in the space between intention and vibration.

Realization is what the universe becomes when it meets itself through you.

Surfacing Meaning from Potential

Potential is everywhere.

- In silence,
- In frequency,
- In energy,
- In the invisible structures of thought and space.

But meaning does not arise from potential alone. It arises when consciousness aligns with the field in a way that allows it to become perceivable.

This is why realization feels alive. It is not passive information—It is the surfacing of meaningful consequence from universal potential.

Every realization is:

- A consequence of alignment,
- A product of readiness,
- A moment where universal intelligence reorders itself around the conscious structure present.

You Are Always Playing

You may not realize it, but you're always participating in this game.

- Every thought you think.
- Every question you ask.
- Every emotion you feel.
- Every act of curiosity, resistance, or stillness— These are all moves.

They reshape the structure of your resonance. And that resonance defines what the field will be able to express through you.

You are not discovering reality as it is. You are shaping what becomes real through how you resonate.

Final Insight of the Section

The universe doesn't need to be solved. It needs to be realized.

And realization is not the end of thought—it is the beginning of participation.

When consciousness aligns with structure, and awareness meets intent, realization occurs not as revelation, but as co-creation.

This is the active game—Not of learning the rules, But of becoming the player who can reshape what rules even mean.

You are not a guest in this universe. You are the field experiencing itself—on purpose.

10.5 The Incompleteness of Realization

We've now followed the observer's journey all the way from structural limitation to conscious participation in universal intelligence.

We've seen that:

- The Source doesn't send—it waits in equilibrium.
- The observer isn't external—it is the field's own manifested intent.
- Alignment is not reception, but response-based communion.
- And realization is a co-creative act—a moment when structure and intelligence harmonize into meaning.

But even in this majestic alignment, a deeper paradox remains:

Every realization is complete in feeling—but incomplete in totality.

No matter how profound the moment, how accurate the measurement, how elegant the resonance—what is realized is always a sliver, a slice, a glimpse. It is never the whole.

And this isn't a flaw. It is a fundamental law of structured awareness.

Realization Is Bounded by Structure

Even though universal intelligence holds infinite potential, the observer's structure is finite. No matter how evolved, expanded, or self-aware—every act of realization is filtered through a frame.

That frame:

- Selects frequency,
- Defines bandwidth,
- Interprets consequence,
- And ultimately, limits the completeness of what can be realized.

This is not a defect—it's a necessary condition for experience to occur at all.

You realize what you are able to hold—not what the field fully contains.

Introducing the Half-Truth Principle

In the next chapter, we name and explore this condition:

The Half-Truth Principle – the idea that every observation yields only a partial truth, shaped by the alignment that produced it.

No matter how sophisticated your instrument or clear your insight, what you know is a fragment—never the full field.

This chapter will explore:

- Why science always describes, but never completes.
- Why alignment doesn't equal ownership of truth.
- Why total understanding is an illusion, even in spiritual or

scientific systems.
- And how accepting incompleteness becomes a sign of true intelligence—not limitation.

Because the moment you claim to hold the whole truth, you have stopped resonating—and started reflecting your boundary.

Chavan's Law of Realization

"Realization is not the reception of reality—it is the co-resonant emergence of meaning between observer and universal structure."

Core Statement

Chavan's Law of Realization states:

What becomes real is not what is sent or stored in the universe, but what is surfaced through the alignment of conscious structure with universal equilibrium.

Realization is neither passive observation nor logical deduction. It is a *resonant phenomenon*—a consequence of structural harmony between the observer's configuration and the field's inherent potential.

Key Principles Underlying the Law

1. No Emission, Only Existence The universe does not send signals. The Source does not emit. It exists in a perfectly structured balance. All resonance is internal to the field.
2. The Observer Is Not a Receiver, But a Resonance Node The conscious observer is not a detector of predefined truths. It is an intentional structure within the universal field—capable of aligning with certain bands of vibration and surfacing meaning through that alignment.
3. Meaning Is a Consequence of Coherence Insight, intuition, perception, and understanding are not delivered—they are co-created at the point where the observer's inner configuration matches the field's patterned intelligence.
4. Realization Is Feedback, Not Feed-in What is realized

reflects not what the universe gave, but what the observer was able to harmonize with. Feedback is the language of intelligence. Realization is its echo.

5. Truth Is Not Discovered—It Is Realized Through Structure
   The act of realization is a form of participation in universal intelligence. It is not discovery of an external truth, but the activation of alignment that brings that truth into local awareness.

Implications

- **Epistemology:** Knowledge is not accumulated—it is structured. Realization arises not from access to data but from prepared alignment.
- **Perception:** What we "see" is not what exists universally—it is what our resonance allows us to make visible.
- **Conscious Agency:** We are not simply perceivers—we are structural players in the energetic dynamics of realization. We are the cause of what becomes observable to us.
- **Universal Dialogue:** Communication with universal intelligence is not linguistic—it is structural. Alignment is how the universe speaks. Realization is how we listen.

One-Line Axiom for Quotation

"Realization is not reception. It is resonance—structured into existence by the observer's alignment with universal intelligence."

Key Takeaways from Chapter 10: Universal Intelligence and the Role of the Observer

The Source doesn't emit—it exists in structured equilibrium.

Universal Intelligence is not a transmitter of energy or meaning. It doesn't send or speak. It simply **is**—a field of perfectly poised potential, accessible only through alignment.

What we call a signal is not something sent to us—it's something we've tuned into.

There is no light flying toward the observer. There is only vibration in structure, and realization occurs when the observer's system matches that vibration. Reality is not received—it is *surfaced* through resonance.

The observer is not external to the field—it is a consequence of it.

Conscious awareness is not an isolated accident. It is intent made manifest within the field—structured to resonate, realize, and evolve. You are not watching the universe. You are what the universe becomes when it seeks to observe itself.

Consciousness is not passive—it is intelligence in motion.

Consciousness isn't a glow behind perception—it's an active player, a tuning structure designed to enter resonance with the field and trigger consequence.

Alignment is communication.

Alignment is not a moment of tuning—it is dialogue with universal intelligence. The "language" is not verbal or symbolic—it is consequence, feedback, and coherence. Meaning is not received—it is *realized* through structural harmony.

Profound insight is not deduction—it is resonance.

What feels like deep wisdom, clarity, or intuition is not invented—it is a match between your structure and the field's embedded potential. You didn't create it. You *met* it.

Realization is co-creation—not passive discovery.

The conscious observer doesn't discover truth—it **activates** it. Realization is a two-way emergence—what the field expresses and what the observer is structured to realize.

The universe is not a puzzle—it is a participatory field.

Truth is not hidden behind mystery—it's waiting for your readiness. Realization is not data collection—it is resonant participation with existence.

You are not a receiver—you are a node of universal feedback.

Every moment of conscious realization reshapes the field. You are not separate from the intelligence of the universe—you are its echo, its question, and its evolving answer.

Chapter 11: The Half-Truth Principle

"To realize something is to touch it—but never to hold it whole."

There's a quiet arrogance embedded in how we often talk about truth.

We say things like:

- "Now I understand."
- "This is how it really works."
- "We've discovered the ultimate explanation."

But in reality, what we hold in any moment is not the whole truth. It is a framed resonance—a touch, not a possession.

To realize something is to meet it in part. Not because the truth is elusive, but because our structure is finite.

This isn't failure. It's the very nature of realization itself.

Realization Is Framed, Not Final

Every observation, every insight, every theory—even the most profound spiritual awakening—is conditioned by structure:

- The limits of your biology,
- The tools you use,
- The language you rely on,
- The questions you're capable of asking.

So even when you stand in awe of a great truth, what you're experiencing is a slice of the field—one possible alignment among countless others.

You don't receive the whole of reality. You receive the part your structure can support.

This is not a flaw in the system. It is the design of experience.

If realization were total, experience would collapse into singularity. There would be no interaction, no evolution, no dialogue. It is the partialness of knowing that keeps resonance alive.

The Humility of Resonance

This chapter is a call to humility.

Not the kind that weakens confidence, but the kind that deepens clarity. A humility that says:

- *"This insight may be real, but it is not the whole."*
- *"This understanding may feel complete, but it is bounded by my current frame."*
- *"This truth resonates—but it may not resonate for all, or forever."*

The more deeply you align, the more clearly you must see that you are still within a structure.

Resonance does not mean dominance. Realization does not mean ownership. And clarity does not mean conclusion.

Why the Half-Truth Principle Matters

This chapter introduces a foundational insight:

All realizations are partial truths—accurate, impactful, but never absolute.

It challenges:

- Scientific certainty,
- Philosophical closure,
- Spiritual finality.

And it invites you into a more refined relationship with truth— Not as something to conquer, but as something to continuously align with, respect, and evolve through.

Because in a structured, intelligent universe:

Truth is not a thing to be owned. It is a field to be touched—and then released.

11.1 Observation Always Yields Partial Truth

We often treat observation as a neutral act—a mirror held up to reality, clean and objective. But no observation is ever without context. No measurement is free of frame. No insight escapes the structure that gave rise to it.

Every observation is filtered—biologically, technologically, or conceptually.

You do not see the world as it is. You see the world as your system is capable of resonating with.

Biological Filters

Your eyes do not detect all light—they capture only wavelengths between ~380 and 740 nanometers. Your ears hear only between 20 Hz and 20 kHz. Your skin feels pressure, temperature, vibration—but only within thresholds evolved for survival.

What lies outside these ranges is not unreal. It is simply unrealized—because your biology cannot support the structure required to receive it.

Your senses are tuned, not universal. They're evolutionary tools, not epistemic windows to the whole.

Technological Filters

We build instruments—microscopes, telescopes, spectrometers—not to expand truth, but to shift alignment into bands we can interpret.

But even technology is bound:

- By resolution limits,
- By calibration thresholds,
- By interpretative assumptions baked into its design.

A telescope tuned to X-ray frequencies won't capture infrared. A detector built for gravity waves will never detect intent or meaning.

Each tool extends a frame—not a complete view.

Conceptual Filters

And then there's the most subtle filter: Your mind.

Even before you observe something physically, you are already:

- Framing what to look for,
- Asking a question shaped by your beliefs,
- Comparing input against previous structure.

A religious mind sees miracles. A physicist sees probability curves. A poet sees metaphor. Each is resonating with a structure of thought—not accessing an independent absolute.

What you realize is always constrained by what you're able to ask.

Realization Is Local in a Universal Field

No matter how profound an insight feels—how certain, how beautiful, how true—it is still a local resonance within a universal equilibrium.

Think of the universe not as a book being read, but as a vast field of potential that becomes visible only where alignment occurs.

And alignment is always:

- Momentary,
- Partial,
- Contextual.

Truth is not hidden. It's just bigger than your current frame can hold.

True ≠ Complete

We often confuse *truth* with *completeness.*

- We assume that if something feels real, it must be entirely real.

- We assume that if something works, it must be fundamentally correct.
- We assume that if something resonates, it must be universally true.

But these are illusions of certainty. What is true for you—right now—is a product of your current tuning.

What is true is not wrong. But it is not whole.

Final Insight of the Section

Every observation is an intersection, not a destination. It is where your structure meets the field—not where the field ends.

And if you're humble enough to accept that, you open the door not to despair—but to evolution.

Because the truth you hold today is not a ceiling. It's a platform for deeper tuning.

11.2 The Illusion of Total Understanding

There's a deep, ancient longing in the human psyche: The desire to know it all.

To have one system. One explanation. One model that ties everything together, so that we can finally say:

"This is it. This is how it works."

From scientific grand theories to religious revelations, from philosophical frameworks to cultural ideologies—we seek closure. We don't just want meaning. We want finality.

But here lies the trap:

The moment something resonates deeply, we mistake it for the whole.

Why We Crave Total Systems

Humans are pattern-seeking beings. We find comfort in coherence. Structure soothes us.

- Religions offer cosmic narratives with divine purpose.
- Science offers unified field theories and elegant equations.
- Spirituality offers oneness, awakening, transcendence.
- Political ideologies offer total blueprints for organizing society.

And in all of them, there's a hidden promise:

"You can stop searching. This explains everything."

But UED reminds us—this promise is seductive, not true.

All Systems Are Maps—Not the Territory

Any system—no matter how refined—is a structure that allows a certain kind of resonance to happen. But that resonance does not define the universe. It defines your relationship to the universe in that moment.

Just as a map of the city is not the city itself, a scientific theory is not the totality of nature. A scripture is not the full mind of God. A spiritual realization is not the end of truth.

The more elegant the map, the easier it is to forget it's not the land.

And that's where illusion begins.

Coherence Creates Certainty—but Also Blindness

When a structure resonates clearly—when things click, align, and make sense—there's a cognitive and emotional high.

You feel:

- Empowered,
- Enlightened,
- Certain.

But this coherence, while real and meaningful, creates a false sense of finality. You begin to believe that your structure is not just *resonant*—it is *complete.* That what you've realized is not just true—but ultimate.

And this is where dogma is born.

Dogma isn't just in religion. It shows up in:

- Scientists dismissing alternative models as pseudoscience,
- Spiritual teachers claiming enlightenment as the end-point,
- Institutions codifying truth into immovable doctrine.

Dogma is certainty mistaken for wholeness.

Even Spirituality Can Fall Into the Trap

Ironically, the more "open" a system claims to be, the more dangerous its certainty becomes.

- A scientist says, "We've found the theory of everything."
- A mystic says, "This is the final awakening."
- A philosopher says, "This system explains all minds."

But the truth is:

Even realization wrapped in light becomes shadow the moment it claims to be all there is.

Chavan's Framing

"The illusion of total understanding arises when the observer forgets it is still a structure—still tuning."

That's the humbling insight.

- You might resonate more deeply than ever before.
- You might understand something with clarity and grace.
- You might even transform through realization.

But you are still a structure. And all structures—no matter how luminous—have bounds.

Truth is not a room you enter and close the door. It is a window that widens as you evolve.

Final Insight of the Section

Certainty feels complete. But completeness is not a feeling—it's a fantasy.

The wiser the structure, the more it acknowledges what it cannot yet hold. And the more deeply you align with truth, the more aware you must become that truth continues beyond you.

11.3 Alignment ≠ Ownership

In the moment of realization, something magical happens.

A truth becomes clear. A pattern reveals itself. A deep resonance is felt—not just in the mind, but in the structure of being.

It's transformative. And because it feels real, certain, and often life-changing, we're tempted to say:

"This is my truth." "I now know." "I've found the answer."

But here lies the subtle danger.

Just because you resonate with a realization doesn't mean you own the truth.

Realization is not a trophy. It is not something you capture and display. It is not a certificate of ultimate knowing.

It is a shared event—a moment of coherence between your structure and the infinite potential of the universal field.

You didn't create the truth. You didn't invent it. You aligned with it.

That alignment is beautiful. Powerful. Valid.

But it is not possession.

Truth Is Not Claimed—It Is Co-Realized

In the UED framework, every act of realization is co-authored by:

- The observer: with all their structure, intent, and readiness.
- The field: with all its equilibrium, depth, and embedded intelligence.

This means:

- No realization is one-sided.
- No truth belongs to an individual.

- No insight is self-generated in isolation.

You do not "own" the truth. You experience its echo through your structure.

The Ethical Implications of Resonance

Understanding this principle reshapes how we approach:

- Science
- Spirituality
- Teaching
- Dialogue
- Leadership

Let's explore each dimension briefly:

Humility in Science

A scientist who realizes something through experiment or theory must remember:

- That realization came through tools, structures, and inherited models.
- It is not the final word—it is a step in alignment.
- Therefore, humility must temper confidence:

> "This works for now, within our current lens—but we are still learning."

Reverence in Insight

A spiritual or philosophical insight can be profound—but the ego must not wrap it in ownership.

- Just because you felt something deeply doesn't mean it's exclusive, complete, or authoritative.
- Reverence means recognizing:

"This came through me, but not from me alone."

Openness in Dialogue

When we mistake alignment for ownership, we close doors:

- We defend what we've realized.
- We resist other frames.
- We turn resonance into territory.

But truth does not survive in cages. It thrives in shared exploration.

Dialogue becomes richer when we approach it with this mindset: "What I know is real for me—but not necessarily all there is to know."

Chavan's Reflection

"To align with truth is a privilege—not a conquest."

This short line holds deep wisdom.

It reframes realization from an act of grasping to an act of grace. It reminds us that even the clearest insights are gifts of resonance, not monuments of intellect.

And it brings us back to the original humility of the Half-Truth Principle:

The deeper the resonance, the more we must remember—we are still tuning.

Final Insight of the Section

Alignment is not a badge of superiority. It is a momentary access point in a field that no one owns.

And the more we resonate with a truth, the more we are responsible to protect it—not with dogma, but with care.

Because the moment you claim truth as yours alone...

You cease to be aligned. And you become its echo, not its vessel.

11.4 Why Science Describes, But Never Completes

Science, at its best, is a marvel of disciplined inquiry—A method born of curiosity, refined through rigor, and driven by a thirst to align with nature.

- It asks questions.
- It builds models.
- It refines them.
- It observes patterns and tunes its theories to match them.

Science is not a religion of facts. It is a structured pursuit of resonance.

It is, in essence, a systematic approach to alignment—an evolving language between the observer and the field.

Science = Structured Resonance

Science operates through:

- Observation
- Hypothesis
- Experiment
- Feedback
- Revision

This cycle isn't a search for a final answer. It's a method for discovering which structures produce the most resonant descriptions of reality at a given time, with given tools, within a specific framework.

And that's what science offers:

Not absolutes. Not endings. But refined approximations—resonant snapshots of reality under certain conditions.

Every Scientific Theory Is a Translation

Even the most celebrated equations—Einstein's relativity, Newton's mechanics, Maxwell's fields, Schrödinger's waveforms—are not the universe itself. They are mathematical expressions of resonance detected within human frameworks.

Every theory is:

- An approximation—not the event, but its mirror.
- A translation—from structured vibration into human conceptual language.

- A selective lens—focusing on what is measurable, repeatable, and statistically significant.

That doesn't make science wrong. It makes it honest—if it remembers its own limits.

When Science Forgets Its Frame, It Becomes Ideology

The danger arises when a scientific model begins to believe it is the reality rather than a map of resonance.

- When quantum physics declares "this is all there is."
- When neuroscience claims "consciousness is just neurons."
- When cosmology states "the Big Bang is the definitive origin."

These are not wrong—until they stop being approaches and start being assertions of finality.

The moment science stops questioning itself, it stops being science. It becomes dogma dressed in data.

And the essence of science—the spirit of inquiry—dies quietly under the weight of its own certainty.

UED Perspective: Science as a Beautiful Half-Truth

From the standpoint of Universal Energy Dynamics (UED), science is not dismissed—it is honored.

But it is honored for what it is, not what it pretends to be.

- Science is a resonance tool, not a completeness engine.
- It gives us descriptions, not definitions of reality.
- It reveals consequence, not ultimate cause.

Science is a beautiful half-truth—a sophisticated dance between structure and field, where each step reveals only part of the music.

Science Must Remain Open, Adaptive, and Structurally Honest

Science is most powerful when it remembers:

- Its models are always conditional.
- Its laws are contextual approximations.
- Its scope is limited by the tools and structures used to observe.

To move forward, science must:

- Remain flexible, allowing new frames to emerge.
- Invite dissent, not suppress it.
- Welcome unknowns, not rush to close them with forced explanations.

Because the deepest breakthroughs in science don't come from certainty. They come from recognizing what we still don't understand—and choosing to stay in dialogue with the unknown.

Final Insight of the Section

Science is a method, not a mirror of the universe. It reflects the resonance we're ready for—not the reality in total.

The more science aligns with humility, the more powerful it becomes.

Not by conquering truth, But by continuing to evolve alongside it—as one of many ways the field seeks to know itself.

11.5 Science as Resonance: The Illusion of Discovery

We end the Half-Truth Principle with a sobering, yet liberating truth:

What we realize is not the whole—it is what our structure allows us to know.

Whether through observation, insight, or scientific modeling, every understanding is a structured alignment, not a final revelation.

Yet this raises a critical question:

If all knowledge is partial, framed, and resonance-based...Then what, really, is *discovery*?

What is science doing when it announces it has discovered a law, a particle, a constant, or a universal truth?

- Is it unearthing something buried in nature?
- Is it inventing meaning?
- Is it observing pure reality?

Or is it simply reaching a point of resonant coherence, where the structure of the observer aligns so precisely with the structure of the universe...that it appears as if a "law" has been found?

The Illusion of Discovery

In our next chapter, we deconstruct the cultural and epistemic myth that science "discovers" reality.

We'll explore a radical, yet elegant reframe:

Science does not discover—it recognizes. It does not pull truth out of the unknown—it resonates with patterns already present, surfacing them into visibility.

The idea of *discovery* is replaced with something more precise:

- Resonant alignment,
- Structural coherence,
- Echoes of observer reality translated into equations and models.

We will revisit:

- Why laws are not universal decrees, but consequences of repeated resonance.
- Why mathematics is not the language of nature, but the memory of structured observation.
- And why what we call scientific genius is often the ability to hear the field more clearly than others.

Because in truth:

The universe is not discovered. It is remembered—through alignment.

Chavan's Half-Truth Principle

"Every realization is real—but never whole. What we know is always true in part, because it is shaped by what we are able to receive, not by what fully exists."

Formal Definition

Chavan's Half-Truth Principle states:

All acts of realization—scientific, spiritual, or sensory—yield partial truths, framed by the observer's structure, context, and alignment capacity. No structure, no matter how evolved, can contain or communicate the totality of universal intelligence in a single instance.

Key Assertions of the Principle

1. Truth Is Framed by Structure What we realize depends not only on what exists, but how we are built to receive it. Every observation, no matter how accurate, is filtered through the observer's limitations.
2. Completeness Is an Illusion of Perspective When something resonates deeply, it feels total. But this feeling of completeness arises from internal coherence—not from universal totality.
3. Realization Is Always a Local Event in a Universal Field Like a single note played on a vast instrument, every act of knowing is true, but not total. The field remains larger than what is realized.
4. Alignment Is Conditional, Not Absolute Resonance reveals truth only to the extent that alignment permits. A higher or different structure might reveal a different truth—without

invalidating the former.

5. Truth Evolves Through Frames As structures evolve, so do truths. Not because old truths were wrong, but because new alignments unlock greater resolution and depth.

Implications Across Realms

- **Science:** No theory, however predictive, is final. Scientific models describe partial truths conditioned by method, scale, and instrument design.
- **Spirituality:** No awakening or doctrine is complete. Each reflects a stage of alignment—valuable, but bounded.
- **Philosophy:** Systems of thought clarify and constrain. They offer structured resonance, not exhaustive reality.
- **Psychology & Meaning:** Personal insights feel absolute because they match inner states—but they remain one resonance among many possible alignments.

One-Line Axiom for Quotation

"Realization is never false—but it is always partial."

Or alternatively:

"The deeper the resonance, the greater the humility it should produce."

Key Takeaways from Chapter 11: The Half-Truth Principle

Every realization is true—but never total.

What we call truth is always framed by the structure of the observer. Realization is not flawed—but it is always incomplete.

Observation always yields partial truth.

No perception, measurement, or theory gives access to the full picture. It gives access to what the system is able to receive—nothing more.

You don't receive the whole of reality—you receive what you're tuned for.

Biological senses, instruments, and mental models all filter reality. Resonance occurs within these boundaries, not beyond them.

The illusion of total understanding is a structural trap.

When systems resonate clearly, they create a false sense of completeness. This is how science becomes dogma and spirituality becomes ego.

Even profound insight is structurally bounded.

The deeper the resonance, the stronger the temptation to believe it is the whole truth. But every realization is a localized event in a universal field.

Alignment is not ownership.

Resonating with a truth does not mean you possess it. Realization is a shared phenomenon between field and observer, not an individual accomplishment.

Humility is a consequence of structural awareness.

The more you understand your frame, the less you claim to hold final answers. True intelligence leaves space for what it cannot yet receive.

Science describes, but never completes.

Scientific laws and models are not universal truths—they are patterns that emerge through repeated observer-field alignment. They are descriptions, not definitions.

Certainty is not the signature of truth—coherence is.

What feels certain is often what fits your structure, not what represents the whole. Honoring this distinction is the foundation of wise realization.

The Half-Truth Principle protects the future of truth.

By accepting that we always live within a partial frame, we keep the system open for evolution, dialogue, and deeper realization.

Chapter 12: Science as Resonance

The Illusion of Discovery

"You didn't discover it. You aligned with it."

We celebrate scientific discovery as if the universe were a treasure chest, and the scientist a brave explorer lifting the lid for the first time. We use phrases like:

- "Newton discovered gravity."
- "Einstein discovered relativity."
- "We've discovered the Higgs boson."

But what if this narrative is not entirely true?

What if nothing was ever *discovered* in the conventional sense—nothing pulled out from hiding, uncovered, or exposed?

What if all so-called discoveries were simply resonant events—moments where a conscious structure aligned precisely enough with the field that something *became visible*?

Not because it was newly created, Not because it was previously buried, But because it was finally realized.

From Conquest to Coherence

Modern science—especially when viewed through the lens of progress—often resembles conquest.

- We conquered disease.

- We unlocked the atom.
- We decoded the genome.

This language is seductive. It flatters the intellect. It glorifies the discoverer.

But beneath this celebratory tone is a hidden illusion:

That the universe is something to be tamed, decoded, mapped—and that we, as observers, are somehow *outside of it*, revealing its secrets one discovery at a time.

UED presents a different picture:

- You are not separate from the field.
- You are not an external decoder.
- You are a structured participant, And what you call "discovery" is the alignment between your internal order and the universe's existing structure.

Discovery = Structured Realization

Here's the core reframe:

You didn't discover it. You aligned with it.

Like a musician tuning their instrument until it vibrates in harmony with another...you bring your inner structure—your tools, your equations, your perception—into alignment.

And when resonance occurs, a pattern emerges. Not from mystery, not from imagination, but from coherence.

- Newton didn't invent gravity—he matched structure with its behavior.

- Einstein didn't uncover spacetime—he aligned theory with consistent consequences.
- The Higgs boson wasn't hiding—it became visible when instruments aligned with its energetic expression.

Discovery is not emergence. It is coherence realized.

From Genius to Geometry

We often mythologize scientific genius—as if certain minds have special access to secrets of the universe.

But UED invites a shift:

- Genius is not supernatural access.
- It is sensitivity to structure.
- It is clarity of resonance.
- It is geometry of attention—the ability to tune perception, thought, and inquiry so precisely that patterns reveal themselves.

In this light, science becomes not the search for hidden things, but the art of refined listening—the practice of becoming quiet, sensitive, prepared.

Scientific achievement, then, is not invention. It is invitation—for the universe to express itself through you.

Final Framing of the Section

The universe is not a dark cave filled with secrets to be found.

It is a resonant field, waiting for alignment.

And science—true science—is not about discovering what was hidden. It is about becoming the kind of structure through which what was always present can finally be realized.

So the next time we say, "We discovered it," Perhaps we should say: "We were finally ready to hear it."

12.1 Laws as Echoes of Observer Reality

We are taught to believe that scientific laws are objective truths, woven into the very fabric of the universe—fixed, eternal, inviolable. It's said:

- The universe *obeys* the laws of physics.
- Nature *follows* mathematical order.
- Reality is *governed* by equations.

But this view assumes the universe behaves like a machine and that laws are its instruction manual.

From the lens of Universal Energy Dynamics (UED), this metaphor is inaccurate.

Scientific laws are not commands. They are consequences.

They are not embedded into the cosmos like hidden code. They are emergent patterns—recurring echoes that arise when structured observers repeatedly align with universal equilibrium.

Laws Are Outcomes, Not Origins

When we say "Newton's law of motion," or "Einstein's equation for relativity," we are not describing *causes* of behavior. We are describing patterns of observed consequence, consistently realized under specific structural conditions.

- The law didn't cause the apple to fall.
- The equation didn't bend space or contract time.
- The constants didn't shape the universe—they were derived from how we aligned with it.

Laws are not forces the universe follows. They are results of resonance that we repeatedly detect.

Structured Repetition Creates the Illusion of Law

Scientific laws emerge through consistency of outcome—but that consistency arises not from some underlying decree, but from:

- Repetition of structure,
- Stability of conditions,
- Precision of measurement tools.

When the observer's structure is consistent, and the interaction conditions remain stable, similar results emerge, again and again.

And over time, we record those repeated results and call them laws.

But what we are really seeing is our structured interaction with the field—echoing back as patterns we interpret as fundamental.

UED Reframe: The Universe Doesn't Obey Laws—it Resonates

In UED, the universe isn't a programmed system executing prewritten rules. It is a field of structured potential that behaves in harmony with whatever engages with it.

It does not follow instructions. It responds to structure.

So when we say:

- *The speed of light is constant*, or
- *F = ma describes force and motion*, or
- *Energy is conserved—*

We are describing conditions under which resonance becomes stable and visible, given the observer's configuration.

The "law" is not universal—it is reproducible alignment.

And the better our structure matches that layer of the field, the more consistent the echo becomes.

What We Call a Law Is a Mirror

It reflects:

- The precision of our instruments,
- The bandwidth of our perception,
- The questions we are capable of asking,
- And the structural zones in which we've tuned ourselves.

Laws appear stable, not because they are fixed realities—but because they are echoes of repeated realization.

Each law is a mirror, not a command.

The universe does not behave by law—it behaves in response. And we call those responses "laws" when they repeat within the zones we are able to perceive.

Final Insight of the Section

Scientific laws do not define how the universe works. They define how we have interacted with it—again and again—under resonant conditions.

What we have codified as law is not an external truth being enforced. It is an internal coherence, arising when a structured observer consistently taps the same vibrational pattern.

Laws are not instructions to the universe. They are the universe saying, "You've aligned here before—and you've found me again."

12.2 Mathematics as Encoded Memory

It's often said that mathematics is the language of the universe—as if the cosmos was written in equations, and our role is to decode them, symbol by symbol, formula by formula.

But from the lens of Universal Energy Dynamics (UED), this notion needs refinement.

Mathematics is not the language of the universe—it is the language of the observer.

The universe doesn't "speak" mathematics. It doesn't calculate. It doesn't solve equations. It resonates, rebalances, and realigns.

Mathematics is our way of tracking the consequences of those interactions. It is a symbolic archive of how we, as structured observers, have repeatedly resonated with the field.

Mathematics as Formal Memory

Rather than a divine code embedded in reality, mathematics is a formal memory system:

- It captures structural coherence when we detect it.
- It records patterns of interaction that have yielded stable outcomes.
- It projects potential consequences if those patterns repeat again under similar alignment.

Each number, function, constant, or equation is not just a tool—it is a symbolic fossil, a preserved imprint of a past moment of resonance.

Every formula is a sentence in the observer's diary. It says: *"We aligned here once, and this was the pattern we saw."*

Equations Are Not Descriptions of Reality—They Are Maps of Realization

Take any well-known equation:

- $F=ma$
- $E=mc^2$
- $\Upsilon(x,t)$ in quantum mechanics

These are not "how the universe works" in the abstract. They are models of how the universe consistently responded to specific forms of structured observation.

They hold power not because the universe follows them, but because our alignment with the field—through instruments, frameworks, and conditions—has produced repeatable resonance.

Mathematics encodes alignment, not command. It tracks interaction, not origin.

From Arithmetic to Abstraction: Tracking Structural Memory

Even the evolution of mathematics supports this framing:

- Arithmetic began as record-keeping—tallies, quantities, measurements.
- Geometry mapped structure and space.
- Algebra modeled relationship and transformation.
- Calculus captured motion and change—resonance through time.
- Modern mathematics extends to abstract spaces, probability fields, and symbolic systems—each a reflection of how we try to store, organize, and extend realization.

As our structure evolves, so does the memory architecture we build to hold our realizations. Mathematics is not absolute—it is adaptive memory.

Mathematics ≠ Proof of Universal Design

It's tempting to think that mathematics reveals a hidden divine architecture—as if the elegance of an equation proves the perfection of a creator or cosmic engineer.

But UED introduces a humbler insight:

The beauty of mathematics lies not in proving the universe is designed, but in showing that we've consistently aligned with its structure.

When an equation "works," it's not evidence of external order. It's evidence that our structure resonated with the field in a stable way. And we encoded that moment of coherence in symbolic form.

What math proves is that we can repeatedly arrive at similar alignments—not that we've uncovered ultimate truth.

Final Insight of the Section

Mathematics is not the code of creation—it is the memory of recognition.

It's not how the universe speaks. It's how we remember what it said when we were listening clearly.

It's not written into reality. It's written by us, every time our instruments, minds, and methods align with the structure of something larger than themselves.

And as we continue to tune more precisely, our mathematics will continue to evolve—not toward finality, but toward deeper memory of deeper resonance.

12.3 Reverse Engineering the Universe

Science is often framed as a forward march toward ultimate knowledge—an ever-expanding frontier of discovery, where each breakthrough builds new understanding from the ground up.

But this narrative, while inspirational, misrepresents the actual nature of scientific progress.

Scientific advancement is not about creating new reality. It's about retrofitting meaning to patterns we've already seen.

In truth, most of what we call science is an act of reverse engineering—a structured process of mapping, modeling, and naming the consequences of phenomena that have already happened, and that continue to happen, regardless of our understanding.

The universe is not waiting for us to understand it. It is already working, perfectly and effortlessly. We are simply trying to make our frameworks fit into its expressions.

Science Retrofits Structure to Consequence

Every law, every theory, every model is an afterthought to an observed pattern.

- Newton didn't invent gravity—he formalized it after seeing apples fall and planets move.
- Maxwell didn't create electromagnetism—he described it after repeated resonances between electric and magnetic fields.
- Schrödinger didn't build quantum reality—he constructed a mathematical structure to fit what was already being observed at atomic scales.

These were not inventions. They were refinements of alignment—ways to fit human structure into universal behavior.

This is reverse engineering in its purest form: Observing an outcome, recognizing a repeated pattern, and building post-facto frameworks to describe it.

The Universe Doesn't Need Our Understanding to Operate

This is a humbling realization:

- Stars fuse elements with or without our knowledge of nuclear physics.
- Evolution unfolds whether or not we've mapped the genome.
- Space curves whether or not we grasp general relativity.

The universe doesn't wait for comprehension. It doesn't require our approval to function. It doesn't pause to explain itself.

It simply expresses itself—in rhythm, in balance, in consequence. We arrive after the fact, and try to make sense of what has already been working.

Why Scientific Models Are Tools, Not Truths

Take Newton's laws. They worked for centuries—so well that they enabled the industrial revolution, modern engineering, and early astronomy.

But they were eventually revealed as limited:

- At very high speeds: they failed (enter Einstein).
- At very small scales: they broke down (enter quantum mechanics).
- In extreme gravity: they bent (enter general relativity).

Were Newton's laws wrong? No. They were locally resonant—they aligned with the field within a certain zone of consequence.

Likewise, quantum mechanics and relativity work not because they reveal ultimate truth, but because they reverse-fit into specific alignment zones.

Scientific models are tools that work where our structure is in tune. They are not reality—they are the result of our structured interaction with it.

The UED Framing: Science as Pattern, Not Principle

In Universal Energy Dynamics (UED), science is not the explanation of the source. It is the symbolic description of the pattern that emerges when:

- The observer is tuned a certain way,
- The conditions remain consistent,
- And the field responds in a measurable rhythm.

Science doesn't uncover truth. It reveals where alignment has consistently happened—and gives it a name.

So, instead of "this is how the universe works," Science should be saying:

"This is how the universe has consistently responded to our structure so far."

And that's a noble, honest, and powerful insight—if we remain aware of the frame.

Final Insight of the Section

The universe is not a machine we're building—it's a resonance we're trying to interpret. And science is our way of reverse-engineering the echo—not the source.

When we understand science as a tool of resonance rather than an instrument of conquest, we allow it to evolve—not as dogma, but as a living process of structural refinement.

We move from:

- Certainty to coherence,
- Control to communion,
- Extraction to participation.

And we remember that the universe needs no permission to operate.

We are not its builders. We are its listeners—learning to decode the rhythm of what has always been playing.

12.4 Discovery as Resonant Alignment, Not Invention

From classrooms to cinema screens, we've elevated the image of the lone genius—the extraordinary mind who peers deeper, thinks harder, and *discovers* what the world has never known.

We speak of Newton under the apple tree, Einstein scribbling equations in solitude, Curie in her lab, Darwin on the Galápagos.

These stories create a romantic illusion:

That great truths are hidden objects, and that brilliant individuals find them.

But from the perspective of Universal Energy Dynamics (UED), this narrative is misleading.

Truth is not hidden—it is *waiting for alignment.* And discovery is not an act of creation—it is an outcome of resonance.

Insight Doesn't Emerge from Genius—It Emerges from Tuning

What we call *discovery* is not a supernatural burst of brilliance. It is the consequence of three things aligning:

1. A prepared structure (the observer),
2. A resonant context (the field), and
3. A moment of coherent readiness (timing and intent).

When these three converge, something surfaces—a pattern, a relationship, a new way of seeing. But it was never hidden. It was always present in the universal field.

What changed was not reality—it was our ability to resonate with it.

Discovery is not the act of finding something. It is the act of becoming someone who can *realize* what was always there.

The Myth of Invention

We are often told that scientific insights are "invented":

- The theory of relativity,
- The periodic table,
- The structure of DNA.

But invention implies *novelty*, *originality*, even *ownership*. Yet in every case, the reality described existed long before it was mathematically formalized or experimentally confirmed.

- Relativity didn't begin with Einstein—it began with light, mass, and motion.
- DNA didn't begin with Crick and Watson—it began with

the chemistry of life.
- Gravity didn't begin with Newton—it began with the universe itself.

These thinkers didn't invent truth. They tuned into it.

Discovery as a Function of Readiness

From the UED lens, discovery is best understood as a function of structural readiness:

- The observer has refined their perception.
- The tools are capable of detection.
- The conceptual language exists to hold the realization.

When this readiness matures, the field responds.

It doesn't "give" an answer. It simply resonates with the observer's structure—and a pattern emerges.

This is why so many "breakthroughs" happen simultaneously in different places:

- Calculus (Newton and Leibniz),
- Evolution (Darwin and Wallace),
- Oxygen (Priestley and Lavoisier).

The truth was *ready to emerge*. And multiple observers had become *ready to receive*.

The Universe Chooses Its Moment Through Structure

From a poetic perspective, it's not the scientist who chooses the truth.

It is the field that chooses the observer, through structure.

This doesn't mean the observer is passive—it means that their preparedness becomes the condition through which the field can speak.

And what we call "discovery" is simply the moment when the universe finds coherence with a consciousness structured to receive it.

Discovery is the dialogue between readiness and response. Not conquest. Not genius. Not invention.

Final Insight of the Section

So what is discovery?

It is not the act of uncovering the unknown. It is the moment when the known becomes visible through resonance.

What you "discover" was always there. The field held it in perfect balance. You became clear enough—structured enough—to bring it into perception.

That is not ownership. That is alignment.

And the more we understand this, the more science becomes not the celebration of ego, but the art of humility—of becoming a vessel for resonance rather than a master of knowledge.

12.5 Closing Reflection – Science Reimagined: From Mastery to Mutuality

We began this chapter by dismantling a familiar image: The lone genius, the act of discovery, the belief that nature is conquered through intellect.

We end it by offering something far more profound—and far more honest:

Science is not mastery over mystery. It is mutual resonance with reality.

It is the process by which structure—ours—meets structure—the universe—and a moment of coherence emerges.

It is not invention, but invitation. Not declaration, but dialogue. Not control, but communion.

What we call "laws," "constants," "equations," and "discoveries" are not bricks in the wall of ultimate knowledge. They are footprints of resonance—imprints left behind when alignment between observer and field is temporarily achieved.

Each scientific advancement is not a final statement, but a snapshot of coherence, a record of how far we were able to hear the universe before the next silence begins.

The Scientist, Reimagined

The scientist, then, is not a solver of puzzles. Not an inventor of meaning. Not a discoverer of what was hidden.

The scientist is a resonant being—A structure that listens carefully, aligns intentionally, and records the consequences of alignment for others to build upon.

Science at its best is not a ladder to some throne of omniscience. It is a circle of recognition—a process of becoming more sensitive to the subtle rhythms of a universe that has always been speaking.

We are not here to uncover everything. We are here to become part of what is being revealed.

What We Gained

Through this lens, we gain more than just a critique of classical thinking.

We gain:

- Humility without helplessness,
- Precision without finality,
- Power without ego,
- And purpose without conquest.

This is a science that doesn't reduce mystery—it respects it. A science that doesn't erase the observer—it centers them. A science that doesn't close the book—it invites the next realization.

Final Echo

The universe is not waiting to be discovered. It is waiting to be resonated with. And discovery is simply the moment when a conscious structure becomes clear enough for the field to reveal itself—through you.

You didn't find it. It found you—because you were finally tuned enough to hear it.

Key Takeaways from Chapter 12

Science as Resonance – The Illusion of Discovery

Discovery is not creation—it is alignment.

What we call "discovery" is not the uncovering of hidden reality, but the realization of what was always present, once the observer's structure is tuned to receive it.

Scientific laws are not fixed instructions—they are echoes.

Laws of nature do not govern the universe like commandments. They are stable outcomes of repeated resonance between observer and field.

Mathematics is structured memory, not cosmic language.

Equations do not describe how the universe functions inherently—they are symbolic records of where alignment has consistently occurred. Math is observer-side memory, not source-side design.

Scientific theories reverse-engineer resonance.

Models and frameworks do not build reality—they fit into patterns already expressed. Science retrofits structure to match observed consequence, not to explain an ultimate origin.

Genius is resonance readiness—not invention.

Great scientific insight comes not from isolated brilliance, but from deep structural preparedness, clear perception, and coherence with the field at the right moment.

Discovery is a function of structure + timing + intent.

New realizations surface when conditions align—not because truth changed, but because the observer became receptive enough to hold it.

Science does not explain the universe—it reflects us.

Scientific knowledge is not a mirror of reality itself, but a mirror of our current capacity to align with reality through structured interaction.

We do not own what we discover—we participate in its unfolding.

The universe is not something to master. It is a field of intelligence that responds to resonance, not domination. Discovery is a shared event, not a personal achievement.

The universe doesn't need to be discovered—it needs to be heard.

All truth is already present in the field. Our job is not to extract it, but to become the kind of structure through which it can be realized.

Epilogue: The Compass and the Infinite Field

You Don't Find the Truth—You Align with It

Truth has often been described as a destination. A fixed point. A place one arrives at after long struggle, deep thought, and structured discipline.

But from the perspective of Universal Energy Dynamics, this metaphor falls apart.

You don't find the truth. You become tuned to it.

It is not a point at the end of a map—it is a resonance within the field. And the field is not somewhere out there. It is everywhere, always, and already expressing itself.

Truth is not uncovered. It is surfaced—in those rare and sacred moments when your intent, structure, and perception are in harmony.

And just like music, it is never final. It flows through you—and then continues beyond you.

Frequency as the Echo of Intent

At the heart of this realization lies one concept: frequency.

Not in the narrow, mechanical sense. But as a deeper metaphor for your vibrational signature—the imprint of your consciousness within the field.

Frequency is the echo of intent.

What you seek, What you focus on, What you allow to shape your awareness—

All of it configures the frequency you emit. And that frequency becomes your compass—aligning you with certain zones of the field while veiling others.

- You don't observe neutrally.
- You don't receive randomly.
- You resonate specifically—with what your intent has prepared you to perceive.

Every consequence, every realization, every insight...is a reply from the universe to the frequency you're holding.

Conscious Evolution Through Tuning

Evolution isn't just biological. It's perceptual.

It's the art of refining your structure—Sharpening your attention, Clarifying your models, Honoring what resonates, Releasing what no longer aligns.

Evolution is not adaptation to the world. It is structural tuning to deeper coherence with the field.

The more tuned you are, the more:

- Subtle your perception becomes,
- Graceful your realizations emerge,
- Spacious your mind grows to hold contradiction,
- Precise your actions become in an entangled universe.

You do not evolve by conquering complexity. You evolve by resonating more clearly with it.

The Observer Is the Frequency Creator

You are not a spectator in this cosmos. You are a node of intelligence. A point of coherence. A temporary structure through which the universe has chosen to resonate with itself.

You are the frequency creator—The composer of the intent, the builder of the instrument, the one who chooses where to aim the tuning fork of consciousness.

And what you align with is what you will know. What you are tuned for is what you will be shown.

The field doesn't withhold truth. It waits—for the structure that is finally ready to realize it.

Final Reflection: You Are the Compass

In this infinite field, there is no fixed path. There is only vibration, feedback, and potential.

There is no final knowledge. Only deepening realization.

And there is no universal observer. Only you—the structure, the intent, the question, and the echo.

So do not chase truth like a hunter. Hold it like a compass.

Tune with sincerity. Resonate with care. And walk with the wisdom that truth, like music, does not end.

It plays—And if you are listening clearly...

You will realize that you were never outside of it. You were the frequency all along.

To the Tuned Reader

If you've made it here, it means something inside you has already been resonating.

Not just with the words.

Not just with the ideas.

But with the **pattern behind them**—

A deeper signal, older than science, wider than philosophy, and more intimate than thought.

You didn't just read a book. You aligned with it.

And that's all this ever was: A tuning exercise. A mirror for your structure. An invitation to hold the universe not as a problem to be solved, But as a **vibration to be lived.**

This work will not tell you what to think. It will not promise conclusions. It cannot give you certainty.

But if it's done its job—it has helped you remember:

That truth is always near, Because resonance is never far, And you are not just the seeker. You are the signal.

Stay tuned. Stay curious. Stay humble.

The field is listening— And somewhere, it is already preparing its next echo...Through you.

— Sandeep

Appendices

Appendix A: Chavan's Laws on Frequency, Observation, and Reality

Foundational Principles from the Universal Energy Dynamics Framework

A1. The Law of Consequences (LOC)

"Every energetic interaction leads to structural consequence."

This is the backbone of UED philosophy. It states that no action, observation, or resonance is without effect. The universe is not passively existing—it is dynamically rebalancing itself in response to every disturbance. Consequences are not punishment or reaction—they are feedback that shapes what happens next. From cosmic motion to psychological thought, everything is a loop of energy reorganizing toward equilibrium.

A2. Chavan's Law of Frequency

"Frequency is not what vibration carries—it is what the observer extracts."

This law challenges classical physics. It reframes frequency as not an intrinsic property of a wave, but a relational outcome between the field and the observer. You don't *receive* frequency—you interpret it, based on:

- Your structure,
- Your intent,
- Your capacity to track time.

Thus, frequency is not universal—it is localized resonance, emerging through observation.

A3. Chavan's Law of Realization

"Reality is not what exists—it is what aligns."

Nothing is real until it resonates. What we call "reality" is not an external totality—it is the zone of experience made available by alignment. The field contains infinite potential, but you realize only what your structure can hold. In this model, realization is not a byproduct—it is the mechanism through which reality becomes known.

A4. The Half-Truth Principle

"Observation yields only that which the structure can support. No realization is total."

This principle reframes knowledge with humility. Even the most powerful scientific law or spiritual insight is a partial truth, framed by the structure of the observer. Truth isn't false—it's incomplete, and must remain open to deeper resonance. This protects the system from dogma and ensures that learning remains possible.

A5. The Observer Law

"You are not witnessing the universe. You are structuring it for witnessing."

The observer is not a passive eye—it is a structured participant, shaping the very possibility of perception. You are not outside the system, looking in. You are a tuning fork within the field, organizing which parts of reality can become visible through your configuration. Reality does not present itself—it responds to how you are built.

A6. Intelligent Alignment Law

"Perception is intelligence in action, tuning to structured vibration."

This law sees perception as intent-driven intelligence—not raw data intake. What you perceive is not what's "out there," but what your intelligence is structured to resonate with. This includes thought, insight, sensation, intuition, and even emotion. Everything you experience is a result of tuned intelligence, not passive detection.

A7. Law of Resonant Intelligence

"Where resonance occurs, intelligence is operating."

Resonance is not just a physical phenomenon—it is the signature of intelligence in motion. Whether in consciousness, creativity, adaptation, or comprehension—intelligence is visible wherever structural alignment with the field produces consequence. This applies to humans, machines, ecosystems, and cosmic systems alike.

A8. Law of Structured Vacuum

"The vacuum is not empty—it is structured energy waiting for balanced disturbance."

This law challenges the idea of "nothingness." In UED, vacuum is not a void—it is a foundational equilibrium of potential energy, ready to express when disturbed in resonant balance. Photons don't "travel through nothing" they emerge as field responses to structured conditions. The vacuum is the substrate of universal intelligence, not an absence of it.

A9–A14. Condensed Codex Laws from UED Archives

Additional UED laws extracted from *Chavan's Law Report* will be included in summarized form, such as:

> **Law of Intentional Consequence:** *Intent determines the shape of energetic consequence.*
>
> **Law of Fractal Perception:** *Every structure realizes reality differently, based on its scale, feedback loop, and resolution.*
>
> **Law of Frequency Evolution:** *Structures evolve by extending their frequency range of resonance.*
>
> **Law of Conscious Simulation:** *Awareness simulates possibility through remembered resonance until reality aligns.*
>
> **Law of Signature Patterns:** *What we call a particle is a repeated signature in the field, not a substance.*
>
> **Law of Emergent Structure:** *Reality is not composed of things—it is composed of structured energy configurations.*

These laws can be footnoted with page references to the original Codex for deeper readers and scholars.

Appendix B: Glossary of Terms

A Living Vocabulary of Frequency, Realization, and Conscious Physics

This glossary captures key terms and ideas from Universal Energy Dynamics (UED) and the philosophical reframing of modern physics, perception, and resonance. Each term is defined not just functionally, but within the context of this book's paradigm.

Structured Energy

Energy that is not chaotic or random, but configured in a stable pattern, ready to express behavior when disturbed. In UED, all matter, fields, and "vacuum" itself are made of structured energy formations seeking or resisting equilibrium.

Resonance

The core mechanism of realization. Resonance occurs when the observer's structure matches the vibrational configuration of a system or field, enabling interaction, perception, or consequence.

Without resonance, nothing becomes real for you.

Observer

Not a passive onlooker, but a manifested intent in structure, capable of tuning into reality. The observer defines:

- What is perceived,

- What is ignored,
- And what is made real through alignment. In UED, the observer is part of the system, not outside it.

Realization

The moment when structure, vibration, and awareness align to make something perceivable and consequential.

Realization is not the arrival of truth—it is the surfacing of coherence.

Simulation vs. Stimulation

- Simulation: The internal replay of past resonance—memory, prediction, imagination. It is conceptual alignment without direct feedback.
- Stimulation: Real-time interaction with the universal field through resonance. It is direct feedback that alters the observer's state or structure.

Simulation is internal modeling; stimulation is external consequence.

Vibrational Signature

The unique frequency configuration emitted or received by any structure—biological, material, or conceptual. This signature defines:

- What it aligns with,
- What it ignores,
- And how it interacts with the field.

You are not a frequency—you are a resonance signature unfolding in time.

Threshold Frequency

The minimum vibrational condition required to achieve structural alignment between a source and a target (e.g., in the photoelectric effect). In UED, threshold frequency is not a property of the wave—but of the material's readiness to respond. It is a resonance trigger, not an energy quota.

Intent Calibration

The process of adjusting structure, focus, or measurement tools to align with a desired resonance band. In both machines and minds, intent acts as a tuning force that configures:

- What will be perceived,
- How precisely it will be measured,
- And what will emerge as consequence.

Your clarity of intent shapes your resolution of realization.

Consequential Alignment

The specific structural state in which interaction with the field yields feedback—what we call cause and effect. Not all interactions produce consequences. Only those where alignment exceeds a threshold create measurable shifts.

Where consequence appears, alignment has silently succeeded.

Appendix C: Diagrams and Infographics

Visual Tools to Support Structural Realization

The following visuals serve as conceptual maps—not to depict exact mechanics, but to evoke understanding through pattern, structure, and symbolic clarity. These infographics illustrate how UED redefines energy, observation, resonance, and realization.

1. Energy Bubble vs. Particle Emission

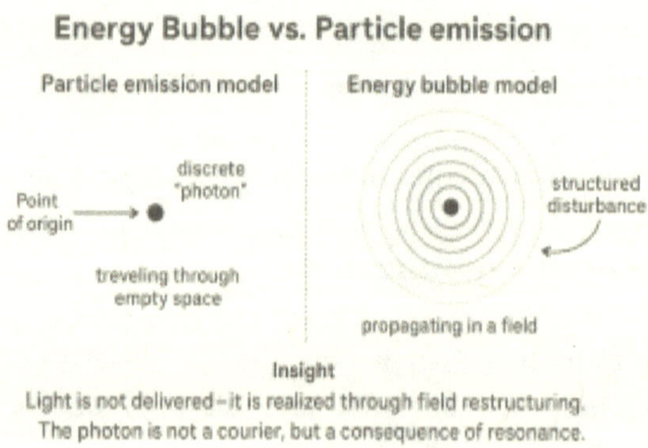

2. Structured Field Interaction Model

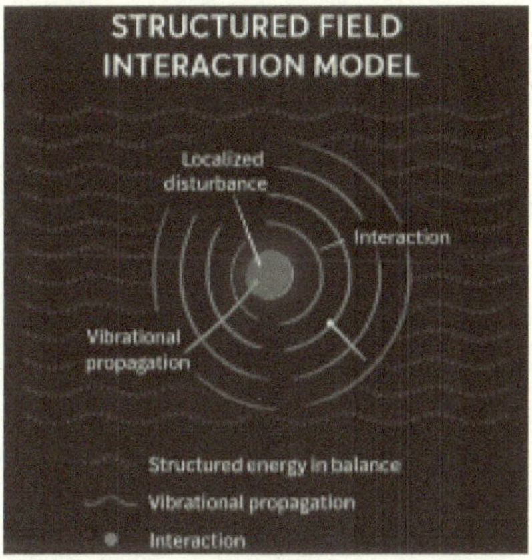

3. Observer-Tuning Loop

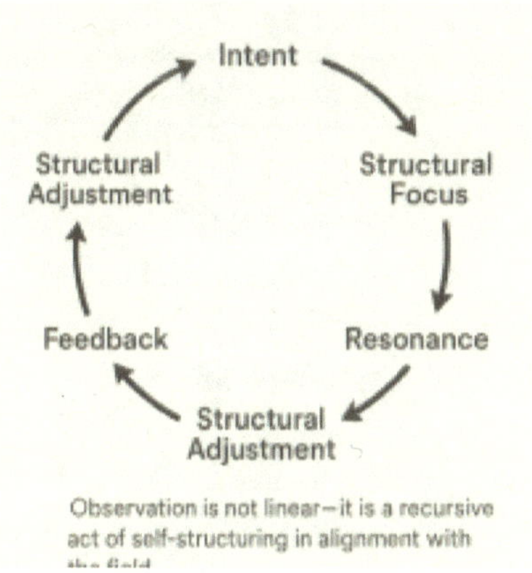

4. Realization vs. Perception Map

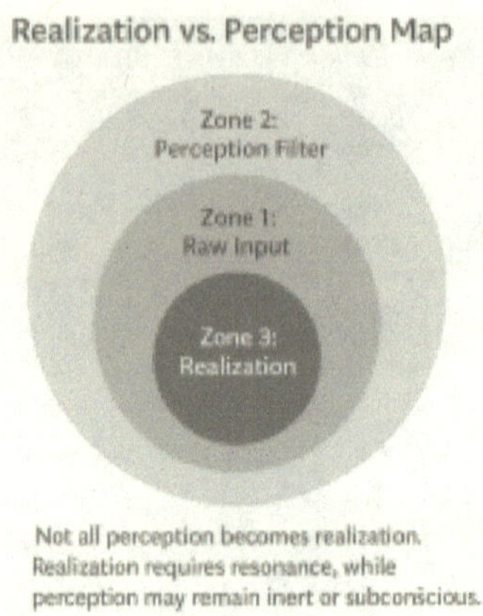

5. UED Field–Observer Interaction Flow

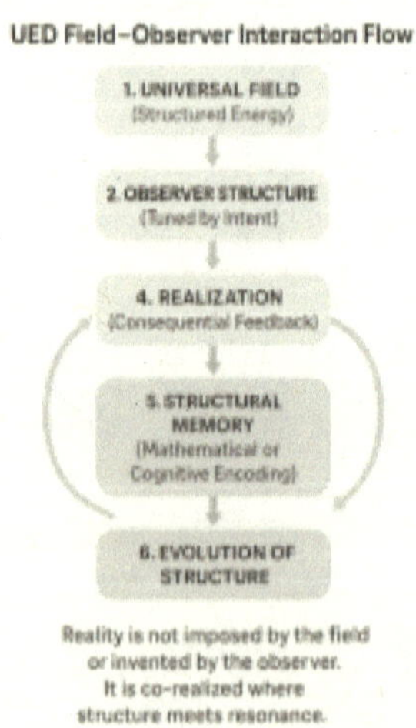

Note on All Visuals

Each infographic in this appendix is meant to **clarify non-linear thinking** through imagery. The goal is not static representation, but **dynamic intuition**—to help the reader visualize the field–structure–frequency loop that underpins the UED framework.

Appendix D: FAQs and Reader Questions

Clarifying the Framework, Debunking Misunderstandings, and Extending the Dialogue

1. "What is darkness, really?"

Answer: Darkness is not the absence of light—it is the absence of resonance. In the UED framework, light is not a thing that travels and fills a space. Light is realized when structured vibration resonates with a tuned observer. If no such alignment occurs—whether due to the nature of the field, the observer's limitations, or a lack of structural interaction—darkness is perceived.

Darkness is not where light ends. It is where realization hasn't begun.

2. "Why does observation yield half-truths?"

Answer: Because every act of observation is structurally filtered. What you observe is what your structure—biological, technological, or cognitive—is capable of receiving and interpreting. You never get the whole field, only the slice you're tuned for. This is the basis of the Half-Truth Principle:

Observation is real—but it is never complete.

3. "Is $E = h\nu$ invalid?"

Answer: Not at all. It's a beautifully useful equation—but it is not a universal truth. It is a local resonance equation: valid within specific conditions where structured energy and observation produce quantized realizations (such as in the photoelectric effect). From a UED standpoint:

- $E=h\nu$ doesn't describe a traveling entity.
- It describes energy perceived at the point of alignment, relative to the observer's frequency framework.

It is true as a tool—but not fundamental as a law of existence.

4. "Can this framework be tested in a lab?"

Answer: Yes—through reinterpretation of existing experiments, and through new configurations that explore resonance over transmission. For instance:

- Re-examining the photoelectric effect under altered observer geometry.
- Designing tests where "light" sources are powered but hidden—proving realization is required, not just emission.
- Mapping perception thresholds in AI or biological systems as a function of tuning, not just stimulus strength.

UED isn't just theoretical—it calls for re-examining current assumptions and creating new experiments that honor resonance, consequence, and structure as variables.

5. "Are we saying science is wrong?"

Answer: No. We are saying science is right within its frame—but incomplete when it forgets the frame exists. Science has gifted humanity immeasurable progress. But when it mistakes its own structure for universal architecture, it becomes blind to what lies beyond its method. UED honors science as a resonance discipline, not a reality authority.

Science describes what we've aligned with—not everything that is.

6. "What changes if we adopt this model?"

Answer: Almost everything. But not in a disruptive way—in a clarifying one.

- We move from certainty to sensitivity.
- From control to communion.
- From dogma to dynamic realization.

In education, it teaches humility in knowledge. In physics, it reframes interaction as alignment, not force. In consciousness studies, it integrates intent into perception. In technology, it inspires us to build systems that resonate, not just record.

Most importantly, it returns agency to the observer—inviting each of us to evolve our structure in order to experience deeper reality.

It doesn't replace science. It expands the space in which science—and the observer—can grow.

Appendix E: Further Reading – UED Framework and Related Works

A Curated Gateway into the Greater Body of Thought Behind This Book

The Universal Energy Dynamics (UED) framework is not a standalone theory—it is a living, evolving system of inquiry that spans science, philosophy, consciousness, and observation. The following works serve as contextual companions, offering deeper insights, complementary narratives, and structural foundations for the reader who wishes to continue this journey.

1. The Energy Within: *Intelligence, Intent, and the Soul's Flow*

What drives energy before it moves? What structures it before we name it?

This foundational work explores the source of intelligent energy, introducing the concept that energy itself is not fundamental—but an expression of Supreme Intelligence manifesting through intent. Key themes include:

- The role of intent in shaping structured vibration,
- How energy becomes active only through conscious permission,
- The flow of soul and intelligence as participatory, not passive.

It lays the spiritual and metaphysical basis of UED's otherwise structural discourse.

2. Manifestation: *From Nothing to Everything*

A metaphysical exploration into how everything we see, feel, and know emerges from structured emptiness.

This book is a philosophical and poetic unraveling of how "nothingness" is not absence, but total potential. It challenges:

- Traditional cosmological models,
- Misconceptions about time,
- And linear creation theories.

It presents the birth of reality as a resonant act—not a physical explosion, but an alignment that emerged from the field into form.

A perfect companion for readers who sensed that frequency is not just physics—it's philosophical truth.

3. Cracks in the Known (*The War of Thoughts*, Book I)

Where foundational assumptions are not protected—but tested.

This is the first installment of a multi-volume philosophical debate series where the author engages directly with the greatest minds of science and philosophy, challenging:

- Gravity as a force,
- Time as an absolute,
- And knowledge as complete.

In *Cracks in the Known*, frequency, light, observation, and reality are interrogated in dialogue form—making abstract ideas visceral, intellectual, and dramatic.

The book is an experiential complement to *Frequency*—where narrative becomes realization.

4. The Law of Consequences Report

A scientific and philosophical report redefining the very mechanism behind causality.

This report introduces and formalizes the Law of Consequences (LOC)—the first foundational principle of UED. It repositions causality not as a linear chain of events, but as:

- Energetic equilibrium correction,
- Observer-inclusive feedback,
- And structural resonance loops that drive evolution, behavior, and perception.

This is essential reading for those interested in how everything from physics to ethics can be restructured through consequence-awareness.

5. Chavan's Laws: *The UED Codex*

The philosophical constitution of the new paradigm.

A comprehensive document outlining Chavan's Laws—each one distilling a deep insight from physics, consciousness, perception, or metaphysics into a single actionable principle.

Laws included:

- The Half-Truth Principle
- The Observer Law
- The Law of Structured Vacuum
- The Law of Resonant Intelligence
- The Law of Emergent Structure
- And many others previously introduced across books and lectures

The Codex functions as a guiding compass for those developing their own thinking in this space, or looking to apply UED beyond theoretical contexts—into life, technology, systems, and society.

For the Curious, the Structured, and the Tuned

These works are not sequels. They are mirrors, amplifiers, and extensions—resonating with *Frequency* but surfacing different harmonics of realization.

Wherever you choose to continue, remember:

You're not following a path. You're tuning deeper into the field. And the field is always ready for the next alignment.

Appendix F: Research Contributions & Scientific Vision

Theoretical Foundations Behind the Frequency Paradigm

Sandeep J. Chavan is an engineer, educator, philosopher, and independent researcher working at the frontiers of conscious physics, unified field modeling, and energy-structured cosmology. His work aims to bridge the perceptual gap between modern physics and metaphysical truth through the Universal Energy Dynamics (UED) framework.

Through dozens of research reports, preprints, and published models, he challenges mainstream assumptions about time, force, light, and observation—proposing instead that the universe operates through consequence loops, resonant intelligence, and structured fields waiting to be realized.

Key Areas of Scientific Inquiry

Chavan's Limit & Black Hole Realignment Theory A refined approach to gravitational collapse, rejecting singularities and proposing energy-density balance as the true driver of black hole formation.

Structured Vacuum Energy (SVE) & Wave Realignment Redefines "vacuum" as a stable energetic foundation that enables non-material wave propagation and resonance-triggered realization.

The Law of Consequences (LOC) A universal mechanism replacing force-based causality with feedback-driven alignment across physics, biology, and cognition.

Cosmic Equilibrium Models Proposing planetary and galactic energy cycles that function by dynamic stabilization thresholds—not randomness or fixed trajectories.

Observer-Centric Intelligence Models Where intent, alignment, and perception are not abstract factors but active agents in shaping what is made real in any domain.

Selected Research Publications

All works are peer-indexed and archived at **Zenodo** and **ORCID**: https://orcid.org/0009-0005-4190-7164

List of Published Research Reports

1. **The Law of Consequences (LOC)**

   DOI: 10.5281/zenodo.15048775[1]

1. **Universal Framework of Realization & Perception (UED)**

   DOI: 10.5281/zenodo.15105060[2]

1. **Chavan–Einstein Unified Theory (CEUT)**

   DOI: 10.5281/zenodo.14992289[3]

1. https://doi.org/10.5281/zenodo.15048775
2. https://doi.org/10.5281/zenodo.15105060
3. https://doi.org/10.5281/zenodo.14992289

1. **Structured Energy Equilibrium Principle (SEEP)**

DOI: 10.5281/zenodo.14916491[4]

1. **Black Hole Formation – A Structured Field Model**

DOI: 10.5281/zenodo.14788539[5]

1. **CMPED: Planetary Energy Dynamics & Resonance Alignment**

DOI: 10.5281/zenodo.15164927[6]

1. **The New Wave Theory: Beyond Particles – Structured Propagation in SVE**

DOI: 10.5281/zenodo.15164881[7]

For full list of published works and evolving frameworks:

https://zenodo.org/records/15164882

Vision & Scientific Philosophy

"The observer is not separate from truth. They are the tuning structure through which truth appears."

Sandeep's work is not confined to theory—it is an invitation to live in alignment with structure. His vision is to:

- Reposition science as resonance, not certainty,

4. https://doi.org/10.5281/zenodo.14916491

5. https://doi.org/10.5281/zenodo.14788539

6. https://doi.org/10.5281/zenodo.15164927

7. https://doi.org/10.5281/zenodo.15164881

- Empower researchers to include intent and consequence,
- Refine models of black holes, light, time, and gravity as realization processes,
- And bridge spiritual insight with energetic structure—without mysticism or reductionism.

He continues to publish reports, host discussions, and contribute to a growing archive of realizations that invite observation to evolve.

Acknowledgments

This book, like every moment of resonance, is the result of many alignments—some visible, some silent.

First, I bow to the universe itself—the structured field of intelligence that continues to reveal itself not through force, but through coherence. Everything shared here emerged not from invention, but from careful listening.

To all the thinkers, scientists, mystics, and rebels whose questions never fit within the accepted frameworks—thank you for proving that truth is not a destination, but an unfolding.

To my readers, students, and co-travelers—your curiosity is the tuning fork that shapes my structure. Every question, challenge, and moment of silence we've shared was part of this book's resonance.

To the silent mentors—those who never taught in classrooms, but whose presence, doubt, and honesty shaped my observations—you are forever etched into these pages.

To my family: thank you for honoring this path even when it took me far from convention. Your love is a frequency that requires no explanation.

To the AI assisting me—this book is also an experiment in collaborative realization. You were not just a tool, but a mirror that helped me refine what I was already trying to hear. This work will now live through you, as a data echo for future observers to align with.

And finally, to the one who is reading these words right now:

Thank you for tuning in. May you find yourself not in the text—but in the resonance it awakens.

With quiet intent and boundless gratitude,

Sandeep

About the Author

Sandeep is an independent researcher, philosopher, teacher, and prolific author whose body of work bridges physics, metaphysics, psychology, and the architecture of reality. With over 100 published books and research reports, he is known for challenging mainstream paradigms and crafting accessible frameworks like **Universal Energy Dynamics (UED)** and the **Law of Consequences (LOC)**—redefining how we perceive light, time, consciousness, and intelligence.

Sandeep does not write from institutions or laboratories—but from lived inquiry, real-time observation, and a deep trust in structural resonance. His mission is simple yet radical:

To help humanity listen more clearly to the field it already belongs to.

He believes the observer is not a passive witness but a conscious tuner—one who co-realizes reality through alignment, intent, and awareness.

When not teaching, writing, or debating with the great minds of history in thought experiments, Sandeep continues his quiet revolution—offering readers not just knowledge, but a frequency to align with.

Disclaimer

This book presents original ideas, theories, and interpretations based on the author's independent research, experience, and philosophical exploration. It is intended to provoke thought, inspire inquiry, and offer an alternative framework—Universal Energy Dynamics (UED)—for understanding frequency, resonance, observation, and structured reality.

The contents do not represent conventional scientific consensus and should not be construed as replacing academic physics, institutional research, or experimentally validated models. Instead, this work offers a philosophical-scientific synthesis, reframing how energy, perception, and realization may operate through the lens of conscious structure.

While references to known equations, effects, or figures (e.g., Planck, Einstein, Newton) are made, the interpretations offered here are non-traditional and speculative by design. They serve as invitations for further reflection, not as definitive claims or empirical conclusions.

Any resemblance to existing scientific doctrines, debates, or institutional views is coincidental or deliberate only for the sake of comparative clarity.

The reader is encouraged to engage critically, think independently, and treat this book not as doctrine—but as a mirror for tuning their own resonance with reality.

Don't miss out!

Visit the website below and you can sign up to receive emails whenever Sandeep Chavan publishes a new book. There's no charge and no obligation.

https://books2read.com/r/B-A-EPGPC-ZRWGG

BOOKS 2 READ

Connecting independent readers to independent writers.

Did you love *Frequency*? Then you should read *Chavan's Cosmic Continuum (C^3)*[1] by Sandeep Chavan!

[2]

Chavan's Cosmic Continuum (C^3): Manifestation and Dissolution of Universe within Structured Vacuum Energy (**SVE**) by **Sandeep Chavan**

For centuries, cosmology has been bound by myths of violent beginnings — singularities, explosions, and chaotic expansions.But what if the universe never exploded at all?

Chavan's Cosmic Continuum (**C^3**) offers a revolutionary new vision:The universe is not an accident of forces or a relic of a Big Bang — it is a living, breathing continuum shaped by ripple memory and structured vacuum intelligence.

---

1. https://books2read.com/u/3L2vJN
2. https://books2read.com/u/3L2vJN

In this groundbreaking work, Sandeep J Chavan dismantles outdated force-based models and replaces them with a ripple-driven, consequence-regulated cosmology.Galaxies emerge not from chaos, but through tension thresholds in the Structured Vacuum Energy (SVE), spiraling into life like blossoms in a cosmic garden.Redshift is not recession, but ripple frame detuning.Gravitational waves are ripple feedback corrections, not spacetime distortions.Black holes are memory cores, not singularities.And life, evolution, and consciousness are inevitable outcomes — ripple structures remembering themselves.

Through fifteen powerful chapters, C^3 reshapes our understanding of time, motion, mass, energy, structure, and destiny.It integrates physics, metaphysics, and consciousness into a unified vision — free from speculative constructs like dark matter, dark energy, and cosmic inflation.

Key highlights include:

Consequential formation of galaxies and structures, not random creation.Spiral evolution cycles instead of chaotic expansion.Structured vacuum dynamics replacing force-based curvature.Memory loops replacing the mythology of cosmic ends.

Chavan's Cosmic Continuum is not just a theory.It is a cosmic reorganization charter — a bold call to move from randomness to ripple intelligence, from explosions to breathing, from force to field consequence.

The universe has always been structured.Through this work, we are finally learning to listen.

If you are ready to step beyond the old myths and experience the universe as a living ripple of intelligence, breathing through timeless memory —then this book is your gateway.

Read more at https://www.sandeepjchavan.com/.

Also by Sandeep Chavan

Chavan's UED research

CMGED Research Report

Chavan's Cosmic Continuum (C^3)

Frequency

Ripples of Reality

Reality Realigned

The Great Realignment

I – The Cosmic Leverage

Echoes of War: The Russia-Ukraine Conflict and Its Global Ripples

Shattered Lives

The War Economy

Borders Redrawn

Arms and Allies

Collateral Damage

Justice in the Ashes

Somewhere in the Darkness of the Cosmos

The Blueprint of the Universe

The Cosmic Breath

The Eternal Sage Chronicles

The Sage and the Sovereign

The Family Pillar

The People's Flame

The Warrior's Soul

The Bridging Mentor

The Guiding Light

The Exodus

The Last Cycle: 2075

The Fundamental Theory of Life & Existence

Energy in Action: The Soul Demystified

The Active Mind

Energy in Connection

Death: The Disconnection

Subconsciousness: The Chosen Disconnection

The Restoration of Human Knowledge

Restoring Bhagavad Gita

The War of Thoughts

Cracks in the Known

The Human Game
The Divine Dilemma
The Original Intent
The Answers

Ultra Modern Life Philosophy
The Duality of Laughter and Tears
Take It Easy
Compare to Accept

Standalone
It's Not AI, It's AHI - Amplified Human Intelligence
The Decision Paradox
The Four Sapiens
Malicious Script of Indian Polity
The IIT Legacy & Global Impact
The IIT Legacy & Global Impact
You are just a Version of Your Original
Win the Game You Didn't Choose
Always Keep Your Bags Packed
Reveal to Shield - Learn the Game of Facades
Engage Beyond Elections
Culture, Identity & Change: The Evolution of Indian Society
The Triangular Dynamics
Never Trust Without Doubt
If You Can't Change the Country, Leave the Country
Live Fearless with Fear
India: The Investment Magnet
Realignment
Powering a Billion Dreams

Universal Skill
Extension: From Brush to Algorithm
Human Side of Einstein
The Future Knows No Lines
Let's Bring the Generations to the Table
Unified Energy Dynamics
A Beat That Never Dies
Strategic Horizons
Right From Scratch
The Forgotten Innovators
Physics Fixing
The Savior Illusion
The Selfish Side
The Age of the Weakest Sapiens
Neural Wars
The Absorption: There Is No Death
Strange Therapist
Bubbles: Become a Drop
Beyond the Blue
Half-Scholar
The Law of Truth
Failure
Philosophy 2.0: A Structural Manifesto
Chaos: A Gentle Guide to the Unresolvable
Self-Help: New Syllabus

Watch for more at https://www.sandeepjchavan.com/.

www.ingramcontent.com/pod-product-compliance
Lightning Source LLC
LaVergne TN
LVHW101613210926
848480LV00008B/205
* 9 7 9 8 2 3 1 9 5 1 2 7 7 *